The Biological
Chemistry of
Magnesium

The Biological Chemistry of Magnesium

EDITED BY

J. A. Cowan

J. A. Cowan
Department of Chemistry
The Ohio State University
Columbus, OH 43210

This book is printed on acid-free paper. ⊗

Library of Congress Cataloging-in-Publication Data

The Biological chemistry of magnesium / edited by James A. Cowan.
    p.   cm.
    Includes bibliographical references and index.
    ISBN 1-56081-627-9
    1. Magnesium—Physiological effect. I. Cowan, J. A. (James A.)
    QP535.M4B54 1995
    574.19′214—dc20

94-26741
CIP

Printed in the United States of America

ISBN 1-56081-627-9 VCH Publishers

Printing History:
10 9 8 7 6 5 4 3 2 1

Published jointly by

VCH Publishers, Inc.
220 East 23rd Street
New York, New York 10010

VCH Verlagsgesellschaft mbH
P.O. Box 10 11 61
69451 Weinheim, Germany

VCH Publishers (UK) Ltd.
8 Wellington Court
Cambridge CB1 1HZ
United Kingdom

To those who have taught and influenced me.

TCL, GJ, AWB, SJT, JKMS, HBG, SML, SGS

Contents

Contributors   *xv*

1. Introduction to the Biological Chemistry of Magnesium Ion   1
J. A. Cowan

1.1. Introduction   1
1.2. Magnesium in the Natural Environment   2
1.3. Magnesium in Living Organisms   5
1.4. Mineral Forms in Vivo   7
1.5. General Physicochemical Properties   8
    1.5.1. Inner vs. Outer Sphere   9
    1.5.2. Complexes with Ligands   11
    1.5.3. Lewis Acidity   12
1.6. Overview of the Biochemical Roles of Mg^{2+} 14
    1.6.1. Structure   14
    1.6.2. Enzyme Activation   15
    1.6.3. Complexes with Nucleic Acids   17
    1.6.4. Cell Walls and Membranes   17
1.7. Thermodynamic Considerations   18
1.8. How to Study the Biological Chemistry of Mg^{2+} 19
1.9. Closing Remarks   21
    References   21

2. Physical Methods for Studying the Biological Chemistry of Magnesium   25

Torbjörn Drakenberg

2.1. Introduction   25
2.2. Detection Techniques   26
    2.2.1. Atomic Absorption   26
    2.2.2. Ion-Selective Electrodes   27
    2.2.3. Nuclear Magnetic Resonance   27
    2.2.4. Magnesium Chelators   30
2.3. Equilibria   32
    2.3.1. Equilibrium Dialysis   32
    2.3.2. Titration with Fluorescent Chelators   32
    2.3.3. NMR   37
    2.3.4. Thermodynamic Measurements   38
2.4. Kinetics   41
    2.4.1. Slow Kinetics   42
    2.4.2. Stopped Flow   42
    2.4.3. NMR   44
2.5. Data Treatment   48
    References   50

3. Metal Substitution as a Probe of the Biological Chemistry of Magnesium Ion   53

Anton Tevelev and J. A. Cowan

3.1. Introduction   53
3.2. Metal Substitution—Spectroscopic Methods   54
    3.2.1. NMR Spectroscopy   54
    3.2.2. EPR Spectroscopy   65
    3.2.3. Electronic Absorption Spectroscopy   70
3.3. Metal Substitution—Chemical Methods   73
    3.3.1. Ion Complexes as Probes of Catalytic Mechanisms   74
    3.3.2. Inert Complexes as Probes of Substrate Conformation   74
3.4. Substrate Modification   75
    3.4.1. Phosphorothioates as Probes of Magnesium Binding   75
    3.4.2. Modification of a Substrate to Distinguish Diastereomers of a Metal–Substrate Complex   75
    3.4.2. Superhyperfine Interactions between Mn^{2+} and ^{17}O 78
3.5. Enzyme Modification   78
    3.5.1. Identification of Metal-Bound Ligands   78

3.6. Closing Remarks   81
References   81

4. Modes and Dynamics of Mg^{2+}–Polynucleotide Interactions   85

Dietmar Porschke

4.1. Introduction   85
4.2. A Short Comparison of Mg^{2+} with Other Ions: Thermodynamics and Kinetics of Complex Formation   86
4.3. Relaxation Methods: Advantages and Limitations   89
4.4. Mg^{2+} Binding to Single-Stranded Polynucleotides   90
4.5. Mg^{2+} Binding to Single-Stranded Oligonucleotides   92
4.6. Mg^{2+} Binding to the tRNA Anticodon Loop   96
4.7. Na^+ Binding to Double-Helical DNA   102
4.8. Discussion   104
References   108

5. Magnesium as the Catalytic Center of RNA Enzymes   111

Drew Smith

5.1. Introduction   111
5.2. Stabilization of RNA Structure   111
    5.2.1. Transfer RNA   112
    5.2.2. Catalytic RNAs   112
5.3. Metal Binding Sites on RNA   114
    5.3.1. Equivalence of Mg^{2+} Binding to Binding of Other Metals   115
    5.3.2. Metal-Binding Sites in Ribozymes—Manganese Rescue   115
    5.3.3. Metal-Catalyzed Hydrolysis of the Phosphate Backbone   116
    5.3.4. Specificity of Binding   117
5.4. Catalysis   119
    5.4.1. Types of Reaction Catalyzed   119
    5.4.2. Participation of Mg^{2+} in Catalysis   120
    5.4.3. Probable Mechanisms; Phosphoryl Transfer Chemistry   121
    5.4.4. Mechanism of Cleavage—5′ Hydroxyl Leaving Group   123
    5.4.5. Metal Ions in Catalysis—5′ Leaving Group   124
    5.4.6. Magnesium-Bound Water as a General Base   124
    5.4.7. Coordination to the Substrate Phosphate   125

5.4.8. Metal Ions in Catalysis—3′ Leaving Group   125
5.4.9. Magnesium-Bound Water as a General Base—
        Activation of the Nucleophile   126
5.4.10. Mg^{2+} as a Lewis Acid   127
5.4.11. Stereochemistry and Ligands   127
5.4.12. Number of Active-Site Mg^{2+} 129
5.4.13. Comparison with Protein Nucleases—A Model for
         Metal Ion Catalysis   129
5.5. Summary   130
      References   132

6. Magnesium-Dependent Enzymes in Nucleic Acid Biochemistry   137

C. B. Black and J. A. Cowan

6.1. Introduction   137
6.2. Polymerase Reactions   137
      6.2.1. DNA Polymerase I (Polymerase Site)   138
      6.2.2. RNA Polymerase   139
6.3. Regulation of Topology   140
      6.3.1. Topoisomerase I   140
      6.3.2. Topoisomerase II (DNA Gyrase)   142
      6.3.3. Helicases   144
6.4. Hydrolysis of the Phosphate Backbone   144
      6.4.1. Ribonuclease H   144
      6.4.2. DNA Polymerase I (3′-5′ Exonuclease Site)   146
      6.4.3. Exonuclease III   147
      6.4.4. Eco RI and Eco RV   149
6.5. Concluding Remarks   154
      References   155

7. Magnesium-Dependent Enzymes in General Metabolism   159

C. B. Black and J. A. Cowan

7.1. Introduction   159
7.2. Glycolytic Enzymes   159
      7.2.1. Xylose Isomerase   160
      7.2.2. Enolase   163
      7.2.3. Pyruvate Kinase   165
7.3. Phosphorylation and Dephosphorylation   167
      7.3.1. Alkaline Phosphatase   167
      7.3.2. Ha-Ras p21   168
      7.3.3. Che Y   169

7.4. Amino Acid Synthesis   170
    7.4.1. Glutamine Synthetase   170
    7.4.2. L-Aspartase   172
7.5. Photosynthesis and TCA Cycles   173
    7.5.1. Ribulose-1,5-bisphosphate Carboxylase   173
    7.5.2. Isocitrate Dehydrogenase   178
7.6. Concluding Remarks   180
    References   181

8. Biological Chemistry of Magnesium Ion with Physiological Metabolites, Nucleic Acids, and Drug Molecules   185

J. A. Cowan

8.1. Introduction   185
8.2. General Metabolites   186
    8.2.1. Phosphates   186
    8.2.2. Sugars   188
8.3. Magnesium-Binding Antibiotics   190
    8.3.1. Tetracycline   191
    8.3.2. Quinolone Antibiotics   191
    8.3.3. Aureolic Acid Antibiotics   193
    8.3.4. Anthracycline Antibiotics   195
8.4. Ionophores   196
    8.4.1. Macrocyclic and Chelating Ethers   196
    8.4.2. Carboxylic Ionophores   199
8.5. Magnesium Binding to Nucleic Acids   199
    8.5.1. Outer-Sphere Coordination   202
    8.5.2. Conformational Switching   203
    8.5.3. Inner-Sphere Coordination   203
    8.5.4. Magnesium Clusters   205
8.6. Concluding Remarks   205
    References   205

9. Genetics and Molecular Biology of Magnesium Transport Systems   211

Ronald L. Smith and Michael E. Maguire

9.1. Introduction   211
9.2. Chemical Properties of Mg^{2+} 212
9.3. Mg^{2+} Transport in Eukaryotic Systems   213
9.4. Mg^{2+} Transport in Prokaryotes   214
    9.4.1. Transport Systems of *Escherichia coli*   214
    9.4.2. Transport Systems of *Salmonella typhimurium*   215

       9.4.3. Transport Properties of the CorA, MgtA, and MgtB
Systems   217
  9.5. Conclusion   227
     References   231

10. Regulation of Cytosolic Magnesium Ion in the Heart   235

A. Romani and A. Scarpa

  10.1. Introduction   235
  10.2. Mg^{2+} Efflux from Perfused Rat Hearts   236
  10.3. Mg^{2+} Movement in Isolated Myocytes   238
  10.4. Stimulation of Myocyte Mg^{2+} Uptake   243
  10.5. Stimulation of Mg^{2+} Release   243
  10.6. Conclusions   248
     References   248

Index   *251*

Contributors

C. B. Black
Evans Laboratory of Chemistry
The Ohio State University
Columbus, Ohio 43210

J. A. Cowan
Evans Laboratory of Chemistry
The Ohio State University
120 West 18th Avenue
Columbus, Ohio 43210

Torbjörn Drakenberg
Department of Physical Chemistry 2
University of Lund
Sweden
and
Chemical Technology
The Technical Research Centre of Finland
Espoo
Finland

Michael E. Maguire
Department of Pharmacology
School of Medicine

Case Western Reserve University
Cleveland, Ohio 44106-4965

Dietmar Porschke
Max Planck Institut für biophysikalische Chemie
37077 Göttingen
Germany

A. Romani
Department of Physiology and Biophysics
Case Western Reserve University
School of Medicine
Cleveland, Ohio 44106-4970

A. Scarpa
Department of Physiology and Biophysics
Case Western Reserve University
School of Medicine
Cleveland, Ohio 44106-4970

Drew Smith
NeXagen, Inc.
2860 Wilderness Place
Boulder, Colorado 80301

Ronald L. Smith
Department of Pharmacology
School of Medicine
Case Western Reserve University
Cleveland, Ohio 44106-4965

A. Tevelev
Evans Laboratory of Chemistry
The Ohio State University
120 West 18th Avenue
Columbus, Ohio 43210

Introduction to the Biological Chemistry of Magnesium Ion

J. A. Cowan

1.1 Introduction

In this work we will review some of the front-line research areas in the field of magnesium biochemistry. Research on the functional role of magnesium ion in a wide variety of biochemical settings has recently flourished, for the most part as a result of the increasing interest of inorganic chemists in understanding the role of this ion as an essential cofactor in nucleic acid biochemistry.[1–3] Physiological chemists have long understood the importance of magnesium in the regulation of cell metabolism[4–6]; however, the ion has lacked the luster or "colorful chemistry" of the transition metal ions that have for so long lured the interest of the coordination chemist. The biological chemistry of magnesium (and the other alkali and alkaline earth metal ions) does, in fact, differ quite markedly from that of the transition elements. This derives from the diverse kinetic, thermodynamic, and redox properties of the latter, and their distinctive coordination geometries and ligand preference.[7] Despite these differences there does occasionally exist sufficient similarity in ionic radii and coordination properties to justify the use of a transition metal ion as a probe of the biological chemistry of Mg^{2+}, Ca^{2+}, Na^+, or K^+. The use of isomorphous replacements reflects a phenomenon commonly encountered in nature, especially in mineralogy, where it is common for a cation to be substituted by trace impurities in a mineral lattice.[8] Previously atomic absorption and the radioactive emissions of the ^{28}Mg isotope have been the principal techniques for monitoring the disposition and abundance of magnesium. Nowadays nuclear magnetic resonance and fluorescent chelators have brought magnesium chemistry within reach of a wider biochemical audience,[9] while the use of metal ion probes have satisfied the chemists' demand for unpaired electrons and optical transitions.

In this chapter we will summarize some of the basic principles of the structural and solution chemistry of divalent magnesium that underlies much of the work to be reported in later chapters, where emphasis will be placed on current research trends in the metallobiochemistry of this ion. Several anecdotal facts concerning the element will also be described. Chapters 2 and 3 present a detailed overview of the physical and chemical methods available for the study of magnesium chemistry. In Chapter 4 the interaction of divalent magnesium with polynucleotides is developed. This leads to a discussion in Chapter 5 of magnesium ion as an essential cofactor in ribozyme reactions, which in turn is further developed in Chapter 6 with an examination of metal activation of key enzymes in nucleic acid biochemistry. Chapter 7 continues the discussion of enzymology with more emphasis on other types of metabolic pathways. In Chapter 8 we examine the coordination chemistry of divalent magnesium with low-molecular-weight biological ligands, substrate molecules, drugs, and the interactions of these magnesium complexes in some instances with larger biological macromolecules. In Chapter 9 we begin a discussion of the physiological chemistry of magnesium with an overview of transport mechanisms. We end with an examination of the role of cytosolic magnesium in mammalian cells, taking a biomedical perspective.

We hope that this volume will present, in a succinct manner, the research tools that can be used to understand the biological chemistry of magnesium. It should also give an appreciation of the diversity of the field. While it would be impossible to cover every facet of magnesium biochemistry in such a short volume, it is hoped that the key areas of interest have been highlighted in a manner that both clearly presents the relevant background and identifies the unresolved questions. We hope that the volume will excite interest in this rapidly advancing area of biochemical research.

1.2 Magnesium in the Natural Environment

Magnesium is one of the most abundant metal ions in the earth's crust and has been recognized since the earliest of times. Indeed the name derives from the Magnesia district in Thessally (Greece) from which magnesia stone, more commonly termed talc [$Mg_3Si_4O_{10}(OH)_2$], was first obtained. Other common magnesium-containing minerals in the earth's crust include dolomite [$MgCa(CO_3)_2$] from the Dolomite range in the Italian Alps, and epsomite ($MgSO_4 \cdot 7H_2O$) from the Epsom district in England. Magnesium is also an abundant component ($\sim 10\%$) of olivine, an igneous rock, and clays (up to 20%).[6,10] Magnesium forms a variety of mineral salts, with carbonates, sulfates, and silicates as the most abundant forms as a result of their low solubility. Several of the more common families are listed in Table 1.1. Elemental magnesium is isolated from these minerals by two major commercial routes. First, calcined dolomite ($MgO \cdot CaO$) is fused with a ferrosilicon alloy, and the magnesium is distilled off under reduced pressure. Alternatively, magnesium is

Table 1.1 COMMON SALTS AND MINERALS CONTAINING DIVALENT MAGNESIUM[a,8]

Orthosilicates		
Mg_2SiO_4	Olivine[b]	$Mg_3Al_2(SiO_4)_3$ garnet
$m[Mg_2SiO_4] \cdot n[Mg(OH)_2]$	Chondrodite[b]	$Mg_3Si_4O_{10}(OH)_2$ talc
$Mg_3Si_2O_5(OH)_4$	Asbestos (chrysotile)	
Salts		
$MgCO_3$	Calcite	$MgKCl_3 \cdot 6H_2O$ carnallite
$MgSO_4 \cdot 7H_2O$	Gypsum	MgF_2
MgO		$K_2Mg_2(SO_4)_3$ langbeinite
Aluminates		
$MgAl_2O_4$	Spinel	

[a] Note that some of these salts and minerals may be hydrated to varying extents.
[b] Olivine may be thought of as a set of close-packed oxygens with silicon in the tetrahedral holes and Mg^{2+} in the octahedral holes. $Mg(OH)_2$ adopts the same structure as Mg_2SiO_4, and so a composite mineral form can result.

produced by reductive electrolysis of either fused $MgCl_2$ ($MgCl_2 + CaCl_2 + NaCl$), or $MgCl_2 \cdot xH_2O$ obtained from sea water.

$$2(MgO \cdot CaO) + FeSi \rightarrow 2Mg + Ca_2S_2O_4 + Fe$$

Since sea water contains an abundant supply of magnesium salts, the commercial importance of this element is likely to grow. Figure 1.1 illustrates the lattice structures of some of these common mineral forms, while the distribution of Mg^{2+} and other common ions over the lithosphere and hydrosphere is illustrated by Figure 1.2 and Table 1.2.

Even a casual inspection of the data in Table 1.2 reveals two important trends. First, the relative abundance of the elements in the lithosphere (earth's crust) reflects, for the most part, the ease of nuclear synthesis in the sun, from which all the terrestial elements are initially derived. The lighter elements tend to be more abundant than the heavier elements, although those that readily form volatile or soluble complexes are depleted as these elements are transferred to the hydrosphere (oceans, seas, rivers, lakes) and atmosphere. The composition of the hydrosphere reflects the solubility of various metal cations in the presence of a variety of simple or complex anions (chloride, sulfate, phosphate, and carbonate). Clearly the distribution of metal ions in living cells closely reflects the concentrations in sea water. These ions were most abundant, and therefore available, and were adopted by the forces of evolution to play the major role in biological chemistry.

Magnesium is a common replacement ion for calcium in a variety of natural minerals. For example, divalent magnesium can substitute for Ca^{2+} in the $CaCO_3$ lattice since $MgCO_3$ is isostructural with $CaCO_3$.[11-13] This mixed mineral is often deposited by invertebrate species. Calcite and aragonite are the two most common mineral forms of $CaCO_3$. They serve a variety of biological roles that include formation of a mineral crystal in the balance organ located in the inner ear of

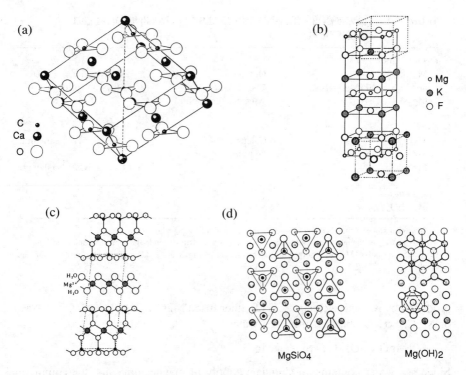

Figure 1.1 Structural forms of magnesium and related calcium minerals. (a) calcite, (b) K_2MgF_4, (c) vermiculite, and (d) $MgSiO_4$ and $Mg(OH)_2$. In (c) and (d) the small black circles represent oxygen. Adapted from Refs. 8 and 11.

mammals (calcite) and fish (aragonite). Substitution of Mg^{2+} for Ca^{2+} is not innoc-uous, and has important implications for crystal formation. In this regard, the solubility products for calcite and aragonite ($K_{sp} \sim 4.7 \times 10^{-9}$ and $6.9 \times 10^{-9} M^2$, respectively) suggest that aragonite is more soluble than calcite. Surprisingly, ar-agonite is found to be the more abundant mineral form in nature. The underlying reason for this concerns the solid state chemistry of magnesium ion. Magnesium does not replace Ca^{2+} readily in aragonite, but does isomorphously replace Ca^{2+} in calcite, and so there is a tendency to precipitate magnesium calcite, according to the relative concentrations of each ion in solution. The local concentration of Mg^{2+} will therefore regulate the relative abundance of calcite versus aragonite in a natural environment.

Magnesium forms numerous other mineral complexes that are of relatively low abundance and dependent on the local environment. For example, during normal cell metabolism lichens release oxalic acid,[11,14] a good chelating agent that can readily complex metal ions available in the immediate environment. This can result in the "weathering" of rock surfaces by chelation of metal ions from the component

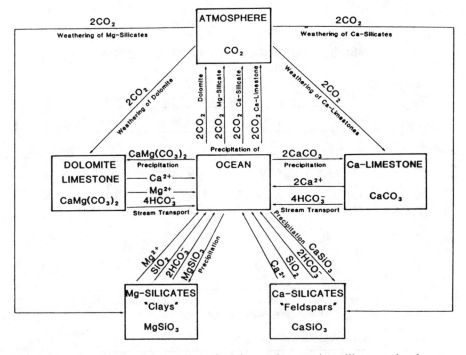

Figure 1.2 Steady-state global cycling of calcium and magnesium silicates and carbonates. Adapted from Garrels, R. M., and Berner, R. A. in Ref. 12.

minerals of the rock face. Magnesium oxalate in its common biomineral form is named glushinskite, and is often found in areas with substantial lichen growth.

1.3 Magnesium in Living Organisms

The relatively high natural abundance of this divalent metal, and many favorable physical and chemical properties (which will be detailed later in this and subsequent chapters), have resulted in the assimilation of Mg^{2+} for many vital biological functions. Magnesium is therefore an essential element, and the average human takes in $\sim$ 0.5 g/day and stores $\sim$ 25 g in mineral forms in bones (65%) and as complexes with nucleic acids and proteins (32%).[15] The availability of magnesium in soils has encouraged the utilization of the element by plants and vegetables. In particular, magnesium is an integral component of chlorophyll [Fig. 1.3 (c)], the light-absorbing pigment that underlies the key light-harvesting reactions of photosynthesis. Although intracellular magnesium is measured at around millimolar levels, it is likely that there is not much "free" Mg^{2+}(aq) in solution, and that most of the ion exists as a complex with cellular ribosomes, membranes, and other charged

Table 1.2   DISTRIBUTION OF METAL IONS IN THE
ENVIRONMENT[a]

(a) Lithosphere[17]

Element	(ppm)	Element	(ppm)
O	460000	F	544
Si	257000	S	340
Al	83000	C	180
Fe	62000	Ci	126
Ca	46600	Ni	99
Mg	27600	Zn	76
Na	22700	Cu	68
K	18400	Co	29
H	1520	N	19
P	1120	Br	2.5
Mn	1060		

[a]   Each element may exist in a variety of oxidation states in the form of
salts, gases, liquids, solutions, or elemental states.

(b) Hydrosphere and living cells

Ion	Seawater (mM)	Blood plasma (mM)
Na^+	470	138
Mg^{2+}	50	1
Ca^{2+}	10	3
K^+	10	4
Cl^-	55	100
HPO_4^{2-}	1×10^{-3}	1
SO_4^{2-}	28	1
Fe^{2+}	1×10^{-4}	2×10^{-2}
Zn^{2+}	1×10^{-4}	2×10^{-2}
Cu^{2+}	1×10^{-3}	1.5×10^{-2}
Co^{2+}	3.1×10^{-6}	2×10^{-3}
Ni^{2+}	1×10^{-6}	0

(c) Approximate bulk elemental composition of a typical 70 kg
human

Oxygen	44 kg	Phosphorus	680 g
Carbon	12.6 kg	Potassium	250 g
Hydrogen	6.6 kg	Chlorine	115 g
Nitrogen	1.8 kg	Sulfur	100 g
Calcium	1.7 kg	Sodium	70 g
		Magnesium	42 g

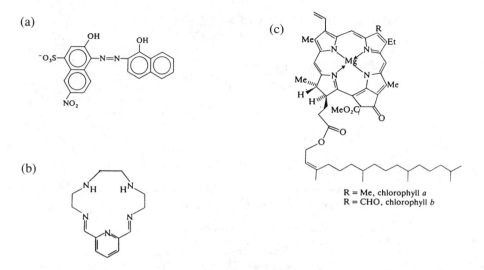

Figure 1.3 Synthetic macrocyclic or chelating ligands. (a) Eriochrome black T, used to quantify Mg^{2+} by a colorimetric test. (b) A Schiff base macrocycle. The X-ray structure of the $Mg(H_2O)_2$ (ligand) Cl_2 complex has been determined.[42] (c) Natural chlorophyll ligands. Note that in (b) and (c) the nitrogen ligand set is unusual for Mg^{2+}.

macromolecules in the cytosol or cell nucleus. A variety of cellular distributions are listed in Table 1.3.

1.4 Mineral Forms in Vivo

Table 1.4 summarizes some of the important biological minerals incorporating Mg^{2+} as either the major or a trace metal ion. We have already noted the importance of certain natural magnesium-containing minerals as biological sensors, but other roles are common. For example, magnesite ($MgCO_3$) is utilized in skeletal formation in many marine creatures (including reef corals), while the bulk (65%) of human magnesium stores are located in bone. Calcium is the major metal ion in skeletal minerals; however, magnesium is commonly found as a structural isomorphous replacement. X-ray diffraction studies demonstrate a close structural relationship between these skeletal minerals and apatites (calcium phosphates),[8,11] and hydroxyapatite [$Ca_5(PO_4)_3OH$] is a reasonable chemical prototype for biological apatites with a hexagonal arrangement of Ca^{2+} and PO_4^{3-} ions around columns of OH^-. In contrast to the carbonate minerals, magnesium ion does not readily substitute for calcium in bone since the ion is small and does not readily incorporate into the apatite lattice structure during crystal growth. The lattice energy for this particular mineral form is large, and so magnesium tends to bind at the mineral surface.[16]

Table 1.3   DISTRIBUTION OF METAL IONS IN A TYPICAL
MAMMALIAN CELL SYSTEM

(a) Intracellular versus extracellular

Ion	$[M^{n+}]_{in}$ (mM)	$[M^{n+}]_{out}$ (mM)
Na^+	10	145
K^+	140	5
Mg^{2+}	30^a	1
Ca^{2+}	1	4
H^+	5×10^{-4}	5×10^{-4}
Cl^-	4	110

[a]   Note that the typical intracellular concentration of "free" Mg^{2+} is ~ 1 mM.

(b) Subcellular distribution of Mg^{2+} in rat liver obtained from three
different determinations

	% of total Mg (1)	% of total Mg (2)	% of total Mg (3)
Nuclei	13.4	16.3	47.8
Mitochondria	21.8	23.2	17.4
Microsomes	48.0	45.2	13.7
Supernatant (cytosol)	12.8	14.1	19.2

Source: Adapted from H. Ebel and T. Günther, *J. Clin. Chem. Clin. Biochem.* **18,**
257–70 (1980). Refer to this article for a detailed discussion of the data.

1.5 General Physicochemical Properties

The choice of specific metal ions for biochemical roles is determined by the relative abundance of an ion and its physicochemical properties. For example, the rapid ligand exchange of Ca^{2+} ($k_{ex} \sim 10^9$ s^{-1}) makes this ion an ideal choice as a secondary messenger relative to Mg^{2+} ($k_{ex} \sim 10^5$ s^{-1}). Clearly an appreciation and understanding of the solution chemistry of Mg^{2+} will provide a great deal of insight on its biological chemistry.

Table 1.5 illustrates how certain key physical parameters of divalent magnesium (ionic radius, solvation energy, transport number, coordination geometry, solvent exchange rates) compare and contrast with adjacent elements in the periodic table. Ions of these elements are usually regarded as "hard"; that is, they typically interact with ligands of high electronegativity and low polarizability (in a biological context this is usually oxygen). Interactions with ligands are often analyzed using an ionic model (based on electrostatic attraction) where association derives from ion–dipole terms. For this reason it was predicted that these ions should bind weakly to neutral ligands, or anions carrying a single negative charge, and generally this is found to be the case (Table 1.6). Strong binding is promoted by chelating or macrocyclic ligands (Fig. 1.3), that is, multiple contacts in a structurally favorable coordination

Table 1.4   BIOLOGICALLY IMPORTANT MINERAL FORMS[a]

Cation	Anion	Formula	Mineral	Distribution and function
Calcium[b]	Carbonate	$CaCO_3$	Calcite Aragonite Vaterite	Exoskeleton in plants, balance sensor in animals, calcium store, eye lens
	Phosphate	$Ca_{10}(PO_4)_6(OH)_2$	Hydroxyapatite	Skeletal, matter, calcium store in shells, bacteria, bones, and teeth
	Oxalate	$Ca(COO)_2 \cdot 2H_2O$	Weddellite	Calcium store in plants
Magnesium	Oxalate	$Mg(COO)_2 \cdot 2H_2O$	Glushinskite	Extracellular mineral
	Carbonate	$MgCO_3$	Magnesite	Skeletal matter in corals

[a] Refer to H. A. Lowenstam and S. Weiner in *Biomineralization and Biological Metal Accumulation* (Eds. P. Westbroek and E. W. de Jong), pp. 191–203, Reidel, for a comprehensive listing of minerals and their distribution in extant organisms.
[b] Divalent magnesium is often incorporated into the lattice structure as a substitute for Ca^{2+}. This may influence the solubility and/or structure of the crystal lattice.

shell. Similar principles are utilized in Mg^{2+} carrier ligands (Fig. 1.4) and binding sites on proteins and nucleic acids (Fig. 1.5).

The relatively large charge density of Mg^{2+} compares well with that of the smaller Li^+ ion (Table 1.5). For this reason, the solution chemistry of Li^+ and Mg^{2+} shows many similarities, and so Li^+ has been used as a chemical probe of Mg^{2+}. This is a manifestation of the so-called diagonal relationships, a similarity in the chemistries of diagonally related elements arising from their comparable charge densities ($Li^+ \equiv Mg^{2+}$, $Be^{2+} \equiv Al^{3+}$, $Na^+ \equiv Ca^{2+}$, $B \equiv Si$).[17] The rather small ionic radius for divalent Mg^{2+} restricts the coordination number to a maximum of six. Coordination of small ligands is preferred (this explains the tendency of Mg^{2+} to retain significant numbers of solvent waters when binding to biological ligands), and a lower coordination number may be adopted if ligands are too bulky. The rather high transport number noted for magnesium in Table 1.5 is also a reflection of the high charge density of Mg^{2+}, which tends to order significant numbers of water molecules in both the primary and secondary solvation spheres. Solvent exchange rates are also seen to follow the trend of charge density; larger monovalent ions undergo exchange more rapidly. It is clear, therefore, that much of the unique chemistry of Mg^{2+} derives from its relatively large charge–radius ratio.

1.5.1 Inner vs. Outer Sphere

There is widespread belief that magnesium biochemistry is dominated by the formation of inner sphere complexes with enzyme and/or substrate. However, a brief review of magnesium solution chemistry and crystal chemistry demonstrates that this should not necessarily be the case.[18] Divalent magnesium has one of the largest hydration radii and charge–radius ratios of any metal ion (Table 1.5). In solution it is heavily solvated. These facts are also reflected in the hydration numbers for

Table 1.5 PHYSICOCHEMICAL PROPERTIES OF THE HYDRATED ALKALI AND ALKALINE EARTH ELEMENTS

Ion	Li^+	Na^+	K^+	Cs^+	Be^{2+}	Mg^{2+}	Ca^{2+}	Ba^{2+}
Radius (Å)	0.73	1.16	1.52	1.88	0.41	0.86	1.14	1.56
Charge density $(q^2/r)^a$	1.37	0.86	0.66	0.53	9.76	4.65	3.51	2.56
ΔH°_{hyd} (kJ mole^{-1})	−515	−405	−321	−263	−2487	−1922	−1592	−1304
Transport number^b	13–22	7–13	4–6	4	—	12–14	8–12	3–5
k_{ex} (H$_2$O)(s^{-1})	5×10^8	8×10^8	10^9	4×10^9	10^2	10^5	3×10^8	10^9
Primary coordination^c	4	6	6–8	8	4	6	6–8	8
Acidity constant (pK_a)	13.9	14.7	—	—	—	11.4	12.6	13.2

^{a} The charge q is taken as the formal valence number (1 or 2).

^{b} Transport numbers estimate the average number of solvent molecules that migrate through solution in close association with an ion. This includes contributions from primary (inner) and secondary (outer) solvation shells, and gives a measure of the electrostatic ordering around each metal ion. The large range of values for each metal represents the degree of uncertainty in this type of measurement; however, the general trends are valid.

^{c} Typically alkali and alkaline earth ions are bound by oxygen ligands (carbonyl, carboxylate, phosphate, and occasionally alcohol or water).

Table 1.6 BINDING AFINITIES OF Mg^{2+} TO COMMON BIOLOGICAL LIGANDS[a]

Ligand	$Log_{10} K$ (electrolyte)	Comments
Acetate	0.51 (0.2M KCl)	Metabolite
Malate	1.55 (0.2M KCl)	Metabolite
Lactate	0.93 (0.2M KCl)	Metabolite
Citrate	3.3 (0.15M $NaNO_3$)	Metabolite
Succinate	1.2 (0.2M KCl)	Metabolite
Glycerol-1-phosphate	1.8 (0.2M KCl)	Metabolite
Glucose-1-phosphate	1.4 (0.1M NaCl)	Metabolite
$ADPH^{2-}$	3.1 (0.1M KCl)	Nucleotide
$ATPH^{3-}$	4.0 (0.1M KCl)	Nucleotide
Aspartate	2.43 (0.1M KCl)	Amino acid
Glutamate	1.9 (0.1M KCl)	Amino acid
Glycine	2.2 (0.09M KCl)	Amino acid
NTA^{3-}	5.4 (0.1M KNO_3)	Chelating ligand
$EDTA^{4-}$	8.7 (0.1M KCl)	Chelating ligand
Eriochrome black R	7.6 (0.1M KNO_3)	Colorimetric agent
Desferri-oxamine B	4.3 (0.1M KNO_3)	Ionophore
SO_4^{2-}	~ 2.0	Solubility product
CO_3^{2-}	~ 7.5	Solubility product
HPO_3^{2-}	~ 24.0	Solubility product

[a] ADP = adenosine diphosphate, ATP = adenosine triphosphate, NTA = nitrilotriacetic acid, EDTA = ethylenediaminetetraacetic acid. All measurements taken at $\sim 25°C$. Data from "Stability Constants," Special Publ. No. 17, Eds. A. Martell and L. G. Sillen, Chemical Society, London, 1964.

crystalline salts of magnesium relative to other alkaline earth metal ions (for example, $MgSO_4 \cdot 7H_2O$, $MgCl_2 \cdot 6H_2O$, $BaCl_2 \cdot H_2O$, $CaCl_2 \cdot 2H_2O$). It might appear that Mg^{2+} has a high affinity for H_2O; however, these data more correctly reflect the difficulty of coordinating large counterions or ligands to the rather small magnesium cation. The question of whether a magnesium cofactor is acting as an inner sphere or an outer sphere complex is important and nontrivial inasmuch as the chemical roles for these two coordination modes are quite distinct[19]; however, the relatively fast exchange kinetics of Mg^{2+} poses a rather significant problem for the elucidation of the appropriate coordination mode in solution.

1.5.2 Complexes with Ligands

We had previously noted that on the basis of an ionic model the alkali and alkaline complexes would not be expected to form relatively stable complexes. Although many isolated cases were known (Fig. 1.3), it was not until 1967, after C. J. Pederson's seminal discovery that macrocyclic polyethers can act as efficient ligands for both the alkali and alkaline earth metals, that systematic and serious investigations of the coordination chemistry of these extremely important biological ions was initiated.[20–22] We now know of course that the coordination chemistry of

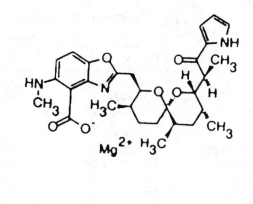

Figure 1.4 Natural magnesium carrier ligands (ionophores). (a) Calcimycin A23187. (b) Lasalocid.

these ions underlies the understanding of a diverse array of biological activity, from ion channels to enzyme activation and membrane structure (Fig. 1.6).

The binding of Mg^{2+} to monodentate ligands tends to be weak (Table 1.6), and chelating environments promote tighter binding. For small molecules the ligand atoms are likely to be separated by only a few intervening atom centers; however, these might be separated by larger distances for structurally more elaborate biological macromolecules (Fig 1.4). Magnesium shows a strong preference for oxygen ligands, especially negatively charged carboxylates, phosphates, or enolate anions. One very important exception is the chlorophyll pigment in photosynthetic cellular membranes. The magnesium ion is bound to the planar tetrapyrrolic pigment [Fig. 1.3(c)] through four nitrogen centers. A few related examples that encourage nitrogen coordination through use of the macrocyclic effect have been documented, for example, the Schiff base macrocycle in Figure 1.3(b).

1.5.3 Lewis Acidity

One aspect of the chemistry of Mg^{2+} that is important for the understanding of enzymatic catalysis lies with its influence on the autoionization of bound H_2O

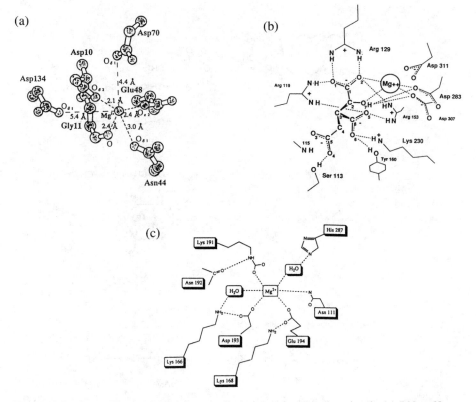

Figure 1.5 Crystallographically characterized magnesium binding sites in (a) RNase H, an endoribonuclease for the RNA strand of RNA·DNA hybrids.[43] (b) Isocitrate dehydrogenase (substrate shown),[44] and (c) Rubisco activation of CO_2 in anabolic biochemistry.[45]

(Table 1.5), that is, the ability of the metal cation either to stabilize a negatively charged anion or polarize a formally neutral functional group such as a carbonyl (C = O) in esters or amides.

$$[ML_5(OH_2)]^{n+} \rightleftharpoons [ML_5(OH)]^{(n-1)+} + H^+$$

$$M^{n+}\text{------}\,^{\delta-}O = C^{\delta+}$$

Transition metal ions may also function as Lewis acids [e.g., Ni^{2+} in urease, Fe^{3+} in acid phosphatases, $(Fe_4S_4)^{2+}$ in hydrolyases], although this is a role that is most commonly filled by magnesium and zinc. The general trend of Lewis acidity follows the order $Ca^{2+} < Mg^{2+} < Mn^{2+} < Fe^{2+} < Co^{2+} < Ni^{2+} < Cu^{2+} > Zn^{2+}$; however, the redox activity of the transition metals makes them less suitable for hydrolysis of sensitive functional groups in proteins and nucleic acids. As a result of their high charge densities, both magnesium and zinc find use as Lewis acid cata-

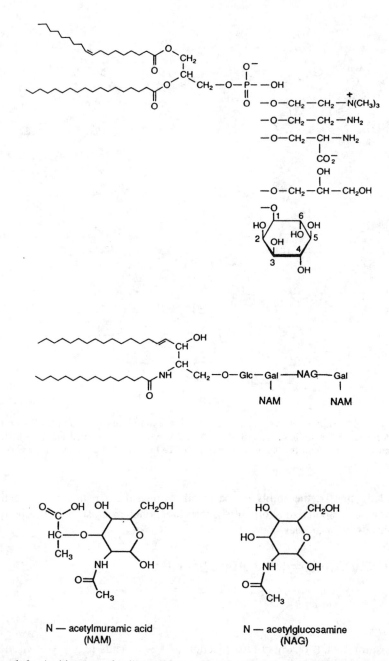

Figure 1.6 Architecture of cell membranes, showing two classes of lipid molecule, phospholipids (above) and gangliosides (middle), with phosphate, carboxylate, and hydroxyl sites for binding divalent magnesium and calcium. The ganglioside lipid class also shows additional functional units attached, including galactose (Gal), glucose (Glc), N-acetylneuraminic acid (NAM) or sialic acid, and N-acetyl glucosamine (NAG) (below).

lysts. Zinc ion is the stronger Lewis acid and is typically employed in the hydrolysis of carbonyl functionality (esters and amides). The harder magnesium ion is more frequently associated with phosphate ester hydrolysis and phosphoryl transfer.

The general mode of action of the magnesium cofactor can be summarized as follows: (1) To stabilize a reaction intermediate: for example, the carbamate derivative formed by RuBisco [Fig. 1.5(c)]. (2) To stabilize a product leaving group: for example, the tight binding metal ion in the putative model for exonuclease activity [Fig. 1.5(b)]. (3) To bind two reactive substrates simultaneously and facilitate reaction through a proximity (template) effect: for example, the weak binding metal ion in the putative model for exonuclease activity [Fig. 1.5(b)].

$$M^{n+} + S \rightarrow M^{n+} - Y \rightarrow M^{n+} + P \qquad (1.1)$$

$$M^{n+} + SX \rightarrow M^{n+} - XS \rightarrow M^{n+}X + S \qquad (1.2)$$

$$M^{n+} + S + X \rightarrow M^{n+} \overset{S}{\underset{X}{\diagup}} \Big) \rightarrow M^{n+} + SX \qquad (1.3)$$

1.6 Overview of the Biochemical Roles of Mg^{2+}

Magnesium is arguably among the most versatile of metal cofactors in cellular biochemistry. The ion serves both intra- and extracellular roles, although the relative concentrations are similar (Table 1.3). In this respect Mg^{2+} differs from the other alkali and alkaline earth ions where substantial transmembrane ion gradients are normal. This reflects the unique biochemical functions of Mg^{2+}. The ion is found in relatively high concentrations (mM); consequently the binding affinities of Mg^{2+} for typical biological ligands (both low molecular weight and macromolecules) are typically low to moderate (10^2–10^5 M^{-1}).[32] Divalent magnesium is the fourth most abundant metal ion in cellular metabolism. Approximately 90% of intracellular Mg^{2+} is bound to ribosomes (complexes of RNA and proteins that mediate protein synthesis) or polynucleotides. Its biological functions include structural stabilization of proteins, nucleic acids, and cell membranes by general surface binding[7]; however, the biochemistry of magnesium ion also includes a stoichiometric requirement to promote specific structural or catalytic activities of proteins, enzymes, or ribozymes.

1.6.1 Structure

Both Table 1.7 and Figure 1.5 demonstrate that magnesium is used to stabilize a variety of protein structures. The chemistry of calcium proteins and enzymes has been more thoroughly developed as a result of the availability of crystallographic data with clear definition of Ca^{2+} binding sites.[23] For example, binding constants

Table 1.7 EXAMPLES OF MAGNESIUM BINDING PROTEINS AND
FUNCTIONAL ROLES

Enzyme	Enzyme function
Eco RI	Restriction nuclease, specifically cleaves ds DNA at G|AATTC sequences
DNA Polymerase I	Incorporates deoxyribonucleotides in the complementary strand of an ss DNA template
(3'-5' exonuclease activity)	Digests ds DNA from the 3' end, releasing 5'-phosphomononucleotides
Topoisomerase I	Relaxes supercoiled DNA by transient cleavage and rejoining of phosphodiester bonds
Che-Y	Phosphatase involved in cell signal transduction, regulatory enzyme catalyzing phosphate transfer reactions as part of the flagellar motor system.
RuBisco	Important enzyme in photsynthesis; Mg^{2+} stabilizes carbamate formation from CO_2, subsequent to carboxylation in the photosynthetic carbon reduction cycle, or oxidation of ribulose-1,5-bisphosphate in photorespiration

for Ca^{2+} vary from the high-affinity ($K_a \geq 10^6$ M^{-1}) regulatory proteins (for example, calmodulin and troponin c) to low-affinity structural or storage proteins such as thrombin, calsequestrin, and phosphodentine ($K_a \sim 10^3$ M^{-1}). In contrast, few magnesium binding sites on enzymes have been crystallographically characterized, although several recent examples have illustrated some structural features (summarized in Fig. 1.5) that may emerge as general motifs.[24] There are many other enzymes where magnesium is known to bind but no role has been identified. For example, Mg^{2+} binds at the interface of the subunits of ribonucleotide reductase and presumably stabilizes the interfacial domain.[25] It is important to note that both polynucleotides and the important enzymes in nucleic acid biochemistry that bind and utilize Mg^{2+} are intracellular species. In contrast, enzymes that have an absolute requirement for Ca^{2+} are typically extracellular digestive enzymes. By discriminating against Mg^{2+}, the cell avoids the problem of premature activation of these degradative enzymes within the cytosol.

1.6.2 Enzyme Activation

Although magnesium and potassium play important roles in the structural biology of nucleic acids,[26–28] they also activate enzymes that regulate the biochemistry of nucleic acids.[29–31] Restriction nucleases, ligases, and topoisomerases are among the many enzymes that are most effectively stimulated by divalent magnesium (Table 1.7). Magnesium plays a role in enzymatic reactions in two general ways. First, an enzyme may bind the magnesium–substrate complex. In this case the enzyme interacts principally with the substrate and shows little, or at best weak, interaction with Mg^{2+} [e.g., MgATP (kinases), Mg isocitrate (isocitrate lyase)].[18] Alternatively, Mg^{2+} binds directly to the enzyme and alters its structure and/or serves a catalytic role.[18,33] Examples of this class include ribonuclease H, exonuclease, and topoisomerase. Although other divalent metal ions may also activate these enzymes, this is frequently accompanied by a reduction of enzyme efficiency and/or substrate specificity. Magnesium binds weakly to proteins and enzymes ($K_a \leq 10^5 \, M^{-1}$), and so magnesium-activated enzymes are not necessarily isolated in the metal-bound form. Magnesium must be added as an essential cofactor to enzyme buffer solutions for in vitro reactions, while in vivo the enzyme is stimulated by the background magnesium concentrations that are of the order of several millimolar.

1.6.3 Complexes with Nucleic Acids

The negatively charged ribose–phosphate backbone has an affinity for metal ions ($K_a \sim 10^2$–$10^4 \, M^{-1}$),[34,35] and it might be expected that the solution chemistry of protein–nucleic acid or ligand–nucleic acid interactions would be intrinsically dependent on the chemistry of the bound counterions. Since the intracellular concentrations of Na^+ and Ca^{2+} are low (Table 1.3), the metal binding chemistry of the nucleic acids in vivo is dominated by K^+ and Mg^{2+}. Ribosomes contain a major fraction of these ions, which are bound to ribosomal RNA and proteins. Magnesium is a key component of the ribosomal machinery that translates the genetic information encoded by mRNA into polypeptide structures. When magnesium concentrations drop below 10 mM, the ribosome dissociates with release of the component rRNA and ribosomal proteins.[36] As a result of its higher charge, divalent magnesium binds to nucleic acids with higher affinity than potassium. The binding of positively charged counterions by nucleic acids is a natural consequence of the polyanionic sugar–phosphate backbone. Metal ions alleviate electrostatic repulsion between phosphates, thereby stabilizing base pairing and base stacking. This results in an increase in melting temperature (T_m) and hypochromism, respectively (Fig. 1.7). The stabilization of nucleic acid structure by the alkali and alkaline earth ions is of great importance, but is poorly characterized. This topic is developed in some detail in Chapters 4, 5, 6, and 8.

1.6.4 Cell Walls and Membranes

The walls of both prokaryotic and eukaryotic cells are constructed from lipid membranes. The latter possess internal structures or compartments (mitochondrion, Golgi apparatus, peroxisomes, etc.) that isolate the various functions essential to normal cell metabolism. The membranes that both envelope these inner compartments and form the outer walls of the cell are composed of a variety of protein polysaccharide and lipid molecules (Fig. 1.6). The peptidoglycan layer found in the outer walls of bacterial cells is composed of a disaccharide repeat unit made up of N-acetylglucosamine (NAG) and N-acetyl-muramic acid (NAM). Figure 1.6 shows the structures of these units and how they are connected by short peptide chains made up from D-amino acids. The outer layer of gram-positive bacteria also contains teichoic acid polymers, of which there are several structural types. Many of

(a)   (b)

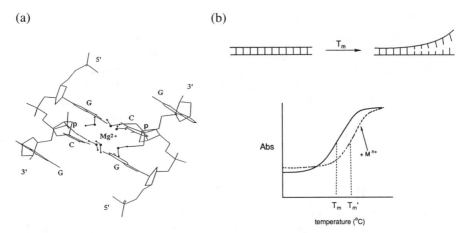

Figure 1.7 (a) Magnesium binding to Z-DNA. Hydrogen bonds from metal-bound H_2O are indicated by dashed lines. (b) Influence of metal ions on the stability of double-strand nucleic acids, resulting from an increase in $\pi-\pi$ interactions through closer base–base stacking, promoted by electrostatic stabilization of the phosphate backbone by Mg^{2+} and K^+.

these are based on phosphodiesters of glycerol with alternating NAG and D-alanyl units on the central hydroxyl. One end of the polymer is attached to a lipid that forms part of a contiguous lipid membrane. It should be clear that many of these surface polymers are polyanionic carboxylates or phosphates. Both magnesium and calcium ions stabilize biological membranes by charge neutralization after cross-linking the carboxylated and phosphorylated head groups of lipids. Chelating agents that bind these metals strongly (e.g., $EDTA^{4-}$) will disrupt cell membranes. Metal ion binding raises the temperature of the phase transition and becomes more prominent with increasing charge on the cation.[37] In this context metal ions also regulate endo- and exocytosis (the processes by which a cell takes in or releases material by occlusion within the cell membrane) by influencing the fluidity and stability of membranes.[38] The bound cations also lower the permeability of the membrane toward water, aiding in the osmotic regulation of solvent transfer across the membrane,[39] and inhibit chemical degradation. There is evidence for selective binding of Mg^{2+}, Ca^{2+}, Fe^{3+}, and Ni^{2+} at the membrane surface (Table 1.8). Surface binding of potentially toxic trace transition metal ions also forms part of a cell's defence mechanism, while the influence of metal ions on membrane structure has important implications for the function of lipoproteins and ion channels.

1.7 Thermodynamic Considerations

To understand the role of alkali and alkaline earth metals in cell metabolism requires an appreciation of transmembrane ion gradients and the mechanisms that maintain and regulate these gradients. There exists a rather simple theory that allows one to characterize these ion gradients quantitatively in terms of transmembrane potentials

Table 1.8   BINDING OF METAL IONS BY ISOLATED CELL WALLS
AND MEMBRANES[a,b]

Organism	Component	Na+	K+	Mg²⁺	Ca²⁺	Mn²⁺	Fe³⁺
B. subtilis	Wall	2.70	1.94	8.22	0.40	0.80	3.58
B. lichen-iformis	Wall	0.91	0.56	0.40	0.59	0.66	0.76
E. coli AB264	Peptidoglycan	0.290	0.060	0.035	0.038	0.052	0.010
E. coli AB264	Outer membrane	0.081	0.025	0.019	0.020	0.012	0.233
E. coli AB264	Cell envelope	0.042	0.082	2.56	0.035	0.140	0.200

[a]   μmol metal (mg dry weight)⁻¹.
[b]   Beveridge, T. J., and Murray, R. G. E., *J. Bacteriol.* **127**, 1502–18 (1976).

and concentrations.[40,41] This chemiosmotic theory provides insight on mechanisms of neurotransmission, voltage gating, and other forms of ion channel activity.

The electrochemical potential (μ_x) of n moles of a species (x) [Eq. (1.4)] is a measure of its free energy under specified solution conditions [activity ($[a_x]$), electrical potential (Ψ), and pressure (P)], where $\mu_x{}^o$ is the electrochemical potential under ideal conditions, Z_x is the charge on species x, V is the volume, and ($P - P_o$) is the pressure difference relative to $P_o = 1$ atm.

$$\mu_x = \mu_x{}^o + nRT \ln a_x + nZ_x F \Psi + V(P - P_o) \tag{1.4}$$

To a good approximation, constant atmospheric pressure can be assumed for solution studies. Equation (1.5) defines the free energy difference (ΔG) for a solution species x separated by a membrane. Since $\mu_{out}{}^o = \mu_{in}{}^o$, we can also write Eq. (1.6).

$$\Delta G = \mu_{out} - \mu_{in} \tag{1.5}$$

$$\Delta G = nRT \ln ([x]_{out}/[x]_{in}) + nZ_x F \, \Delta\Psi_x \tag{1.6}$$

From Eq. (1.6) it is clear that the movement of charged species across a biological membrane is dependent both on concentration gradients and membrane potentials. This is a key equation for relating transport phenomena to the energy state of the cell (Fig. 1.6). The flow of a charged species is driven by an electrochemical gradient, encompassing both a concentration gradient and a membrane potential. If $\Delta G > 0$, movement into the cell will be favored. If $\Delta G < 0$, the spontaneous direction of flow will be out of the cell. When $\Delta G = 0$ (equilibrium) we obtain the Nernst equation:

$$\Delta\Psi_x = (RT/ZF) \ln[x]_{out}/[x]_{in} \tag{1.7}$$

1.8 How to Study the Biological Chemistry of Mg²⁺

A major problem facing researchers pursuing detailed biochemical studies of the alkali and alkaline earth metals relates to the general spectroscopic silence of these ions. Chapters 2 and 3 overview a variety of physical and spectroscopic methods that have been used to monitor the solution chemistry of divalent magnesium. These include both direct and indirect methods, such as ^1H, ^{31}P, and ^{25}Mg NMR, potentiometry, fluorescence, atomic absorption, and ion-selective electrodes. Although each of the alkali and alkaline earth metals is NMR active in at least one isotopic form, multinuclear NMR spectroscopy is not as useful as EPR or electronic absorption as a probe of coordination chemistry. NMR studies of exchangeable metal ions can provide useful kinetic and thermodynamic data on metal binding, but offer little in the way of structural information. This problem can be circumvented by the use of functionally equivalent transition metals that possess convenient spectroscopic handles to probe and investigate their coordination environment. Table 1.9 summarizes some relevant spectroscopic parameters for Mg^{2+} and probe ions that have proven to be of value in studies of magnesium chemistry. Chapter 3 outlines some chemical approaches to this problem in detail. Of central importance here is the need for a similarity in both the size and chemistry of the probe ion. This, together with a desire to apply physical inorganic methods, leads us to the choice of first row transition metals. For various reasons, some of which will be outlined in the following, none of these ions is ideal. However, with an understanding and appreciation for differences in their chemistries,[7] useful information can be obtained.

On the basis of size, divalent nickel is the best substitute for magnesium ion (Table 1.9), but with a $k_{ex} < 10^4$ s^{-1}, it is clear that Ni^{2+} lacks lability. In those cases where the metal ion is locked in place, and rapid exchange is not essential (for example, the ion may stabilize a particular conformation), this is not likely to be a serious problem. However, if release of a bound ligand is required, nickel may prove to be inactive as a substitute ion. Furthermore, the d^8 Ni^{2+} ion prefers to adopt a square planar coordination geometry.

Divalent manganese is the most widely used probe of magnesium chemistry relative to the other first row transition metal ions, showing the greatest similarity

Table 1.9 PHYSICOCHEMICAL AND SPECTROSCOPIC PROPERTIES OF Mg^{2+} AND COMMON PROBE IONS

Mg^{2+}			Probe ions		
Ionic radius (Å)	Coord. number	Probe ion	Ionic radius (Å)	Coord. number	Useful physico-chemical properties
0.86	6	Mn^{2+}	0.97	6	High-spin d^5, EPR, relaxation agent in NMR, similar chemical activity
		Ni^{2+}	0.65–0.85	4–6	Square planar/octahedral diamagnetic low spin
		Li^+	0.73–0.90	4–6	^7Li, $I = 3/2$, 92.6% nat. abun.

with the chemical properties of magnesium ion in terms of its ligand preference and geometry, exchange rates, and a propensity for both inner and outer sphere complexation. In many respects the biological chemistry of Mg^{2+} and Mn^{2+} are closely related, and it is likely that Nature's choice of Mg^{2+} over Mn^{2+} reflects the relative abundances of each ion in vivo ($[Mg^{2+}] \sim 10^{-3}$ M; $[Mn^{2+}] \sim 10^{-8}$ M). Many magnesium-dependent enzymes are also activated by manganese, while manganese transport systems appear to act as backups to magnesium transport. For these reasons Mn^{2+} is often used as a substitute for Mg^{2+}. The ions have similar exchange rates and coordination chemistry; however, the manganous ion is significantly larger in ionic radius (intermediate between Ca^{2+} and Mg^{2+}, Table 1.9), which can lead to differences in the biochemical behavior of the ions. Calcium ion is typically a poor replacement for divalent magnesium.

Consideration of charge densities might lead one to choose Li^+ as a probe for magnesium binding sites. Lithium ion is in fact used in the treatment of various neurological and psychiatric disorders (such as acute depression), and its functional mechanism may involve the competitive binding of Li^+ at Mg^{2+} sites.

1.9 Closing Remarks

In this chapter we have provided an overview of the general coordination and solution properties of magnesium ion as they pertain to its biological chemistry and its scope as an important chemical species in nature. Subsequent chapters will develop these topics in greater depth and detail. Ultimately, however, the selection of any metal cofactor for a specific functional role derives from two main factors: first, the natural abundance and solubility of the metal cation, and second, and more important, its chemical properties.[7] As the reader analyzes the chapters of this book to learn more about the specific functions of divalent magnesium in a variety of biochemical settings and the methods used to obtain information on its chemistry, it is a worthy exercise to view these functions in the context of the fundamental structural, kinetic, and thermodynamic properties of the ion and its complexes.

References

1. (a) Pyle, A. M., *Science* **1993**, *261*, 709. (b) Piccirilli, J. A.; McConnell, T. S.; Zang, A. J.; Noller, H. F.; Cech, T. R., *Science* **1992**, *265*, 1420.

2. (a) Kim, S.-H. *Topics in Molecular and Structural Biology* **1981**, *1*, 83–112. (b) Ryan, P. C.; Draper, D. E., *Biochemistry* **1989**, *28*, 9949–56. (c) Cech, T. R.; Bass, B. L., *Ann. Rev. Biochem.* **1986**, *55*, 599–629. (d) Moore, P. B., *Nature* **1988**, *331*, 223–27.

3. (a) Reid, S. S.; Cowan, J. A., *Biochemistry* **1990**, *29*, 6025–32, and references therein. (b) A number of relevant articles are included in *Adv. Inorg. Biochem.* **1981**, *3*, 103 (Eds.) Eichorn, G. L.; Marzilli, L. A. A good summary can be found in "Metals and Micro-organisms" by Hughes M. N.; Poole, R. K.; Chapman and Hall, New York 1989. (c) Saenger, W., *Principles of Nucleic Acid Structure*, Springer-Verlag, New York, 1984, pp. 290–93.

4. Wacker, W. E. C., *Magnesium and Man,* Harvard University Press, Cambridge, 1980.

5. Williams, R. J. P., *Chem. Soc. Quart. Rev.* **1970**, 331.

6. *Metal Ions in Biological Systems,* Vol. 29, *Compendium on Magnesium and its Role in Biology, Nutrition, and Physiology,* Sigel, H.; Sigel, A. (Eds.), Marcel Dekker, Inc., New York, 1992.

7. (a) Cowan, J. A., *Inorganic Biochemistry. An Introduction,* VCH, New York, 1993. (b) Burgess, J., *Ions in Solution: Basic Principles of Chemical Interaction,* Ellis Horwood, Ltd.; New York, 1988.

8. Wells, A. F., *Structural Inorganic Chemistry,* 4th Ed., Clarendon Press, Oxford, 1975.

9. (a) Drakenberg, T.; Forsen, S.; Lilja, H., *J. Magn. Reson.* **1983,** *53,* 412–22. (b) Lindman, B.; Forsen, S.; Lilja, H., *Chem. Scr.* **1977,** *11,* 91. (c) Forsen, S.; Lindman, B. *Annual Reports on NMR Spectroscopy,* **11A,** 183, 1981.

10. (a) Schachtschabel, P.; Blume, H.-P.; Hartge, K.-H.; Schwertmann, U., *Lehrbuch der Bodenkunde,* Enke Verlag, Stuttgart, 1982. (b) Mengel, K.; Kirkby, E. A., *Principles of Plant Nutrition,* Int. Potash Inst., Bern, 1987.

11. Simkiss, K.; Wilbur, K. M., *Biomineralization. Cell Biology and Mineral Deposition,* Academic Press, New York, 1989.

12. *Biomineralization and Biological Metal Accumulation,* (Eds.) Westbroek, P.; deJong, E. W.; D. Reidel Publishing Co., Boston, 1983.

13. Lowenstam, H. A.; Weiner, S., *On Biomineralization,* Oxford University Press, New York, 1989.

14. Jones, D.; Wilson, M. J.; McHardy, W. J., *J. Microsc.* **1981,** *124,* 95–104.

15. Chapters 4 and 5 in Ref. 6.

16. *Metals in Biology* (Ed. Spiro, T.), *Calcium,* Vol. 6, Wiley-Interscience, New York, 1984.

17. Greenwood, N. N.; Earnshaw, A., *Chemistry of the Elements,* Pergamon, New York, 1984.

18. Black, C. B.; Huang, H-W.; Cowan, J. A., *Coord. Chem. Rev.,* 1994 (in press).

19. Cowan, J. A., *Comments on Inorganic Chemistry* **1992,** *13,* 293–312.

20. (a) Kluger, R.; Wasserstein, P., *Biochemistry* **1972,** *11,* 1544–46. (b) Kluger, R.; Wasserstein, P.; Nakaoka, K., *J. Am. Chem. Soc.* **1975,** *97,* 4298.

21. (a) Bock, R. M., in *The Enzymes,* Boyer, P. D.; Lardy, H.; Kyrback, K. (Eds.), 1960, pp. 3–38. Phillips, R., *Chem. Soc. Rev.* **1966,** *66,* 501–27. (b) Izatt, R. M.; Christensen, J. J.; Rytting, J. H., *Chem. Rev.* **1971,** *71,* 439–81.

22. Cowan, J. A., *Inorg. Chem.* **1991,** *30,* 2740–47.

23. (a) Drakenberg, T.; Forsen, S.; Lilja, H., *J. Magn. Reson.* **1983,** *53,* 412–22. (b) Lindman, B.; Forsen, S.; Lilja, H., *Chem. Scr.* **1977,** *11,* 91. (c) Linse, S.; Johansson, C.; Brodin, P.; Grundstrom, T.; Drakenberg, T.; Forsen, S., *Biochemistry* **1991,** *30,* 154–62. (d) Martin, S. R.; Linse, S.; Johansson, C.; Bayley, P. M.; Forsen, S., *Biochemistry* **1990,** *29,* 4188–93. (e) *Metal Ions in Biological Systems,* Vol. 17 (Ed. Sigel, H.), Marcel Dekker, New York, 1984.

24. (a) Kim, E. E.; Wyckoff, H. W., *J. Mol. Biol.* **1991,** *218,* 449–64. (b) Pai, E. F.; Krengel, U.; Petsko, G. A.; Goody, R. S.; Kabsch, W.; Wittinghofer, A., *EMBO J.* **1990,** *9,* 2351–59. (c) Chevrier, B.; Dock, A. C.; Hartmann, B.; Leng, M.; Moras, D.; Thuong, M. T.; Westhof, E., *J. Mol. Biol.* **1986,** *188,* 707–19. (d) Farber, G. K.; Glasfeld, A.; Tiraby, G.; Ringe, D.; Petsko, G. A., *Biochemistry* **1989,** *28,* 7289–97. (e) Beese, L. S.; Steitz, T. A., *EMBO J.* **1991,** *10,* 25–33.

25. Nordlund, P.; Sjoberg, B. M.; Eklund, H., *Nature* **1990,** *345,* 593–98.

26. (a) Ho, P. S.; Frederick, C. A.; Quigley, G. J.; van der Marel, G. A.; van Boom, J. H.; Wang, A. H.-J.; Rich, A., *EMBO J.* **1985,** *4,* 3617–23. (b) Gessner, R. V.; Quigley, G. J.; Wang, A. H.-J.; van der Marel, G. A.; van Boom, J. H.; Rich, A., *Biochemistry* **1985,** *24,* 237–40. (c) Quigley, G. J.; Teeter, M. M.; Rich, A., *Proc. Nat. Acad. Sci. USA* **1978,** *75,* 64–68.

27. (a) Gessner, R. V.; Frederick, C. A.; Quigley, G. J.; Rich, A.; Wang, A. H.-J., *J. Biol. Chem.* **1989,** *264,* 7921–35. (b) Wang, A. H.-J.; Hakoshima, T.; van der Marel, G. A.; van Boom, J. H.; Rich, A., *Cell* **1984,** *37,* 321–31.

28. (a) Quintana, J. R.; Grzeskowiak, K.; Yanagi, K.; Dickerson, R. E., *J. Mol. Biol.* **1992,** *225,* 379–95. (b) Sundquist, W. I.; Klug, A., *Nature* **1989,** *342,* 825–29. (c) Jack, A.; Ladner, J. E.; Rhodes, D.; Brown, R. S.; Klug, A., *J. Mol. Biol.* **1977,** *111,* 315–28.

29. Steitz, T. A.; Steitz, J. A., *Proc. Natl. Acad. Sci. USA* **1993,** *90,* 16498–502.

30. (a) Needham, J. V.; Chen, T. Y.; Falke, J. J., *Biochemistry* **1993,** *32,* 3363–67. (b) Janeway, C. M. L.; Xu, X.; Murphy, J. E.; Chaidaroglou, A.; Kantrowitz, E. R., *Biochemistry* **1993,** *32,* 1601–9, and leading references in each article.

31. (a) Katayanagi, K.; Miyagawa, M.; Matsushima, M.; Ishikawa, M.; Kanaya, S.; Ikehara, M.; Matsuzaki, T.; Morikawa, K., *Nature* **1990,** *347,* 306–9. (b) Yang, W.; Hendrickson, W. A.; Crouch, R. J.; Satow, Y., *Science* **1990,** *249,* 1398.

32. Huang, H-W.; Cowan, J. A., *Eur. J. Biochem.* **1994,** *219,* 253.

33. Beese, L. S.; Steitz, T. A., *EMBO J.* **1991,** *10,* 25.

34. (a) Manning, G. S., *Acc. Chem. Res.* **1979,** *12,* 443–49. (b) Manning, G. S., *Q. Rev. Biophys.* **1978,** *11,* 179–256. (b) Record, M. T.; Anderson, C. F.; Lohman, T. M., *Q. Rev. Biophys.* **1978,** *11,* 103–78.

35. Black, C. B.; Cowan, J. A., *J. Am. Chem. Soc.* **1994,** *116,* 1174.

36. White, J. P.; Cantor, C. R., *J. Mol. Biol.* **1971,** *58,* 397–400.

37. (a) Triggle, D. J., *Prog. Surf. Membr. Sci.* **1972,** *5,* 267–322. (b) Beveridge, T. J.; Murray, R. G. E., *J. Bacteriol.* **1976,** *127,* 1502–18.

38. (a) Hauser, H.; Levine, B. A.; Williams, R. J. P., *Trends Biochem. Sci.* **1976,** *1,* 278–81. (b) Ochiai, E. I.; Frieden, E. (Eds.), *Biochemistry of the Elements:* Vol. 7, Plenum Press, New York, 1987, p. 205.

39. Agron, P. A.; Busing, W. R., *Acta. Crystallogr.* **1985,** *C 41,* 8–10.

40. Rottenberg, H., *Biochim. Biophys. Acta* **1979,** *549,* 225–53.

41. Westerhoff, H. V.; van Dam, K., *Curr. Topics Bioenergetics* **1979,** *9,* 1–62.

42. Drew, M. G. B.; Othman, A. H-B.; McFall, S. G.; Nelson, S. M., *J. Chem. Soc. Chem. Commun.,* **1975,** 818.

43. Katayanagi, K.; Okumura, M.; Morikawa, K., *Proteins* **1993,** *17,* 337.

44. Hurley, J. H.; Dean, A. M.; Sohl, J. L.; Koshland, D. E.; Stroud, R. M., *Science* **1990,** *249,* 1012.

45. Lundqvist, T.; Schneider, G., *Biochemistry,* **1991,** *30,* 904.

2

Physical Methods for Studying the Biological Chemistry of Magnesium

Torbjörn Drakenberg

2.1 Introduction

Magnesium and calcium are among the most important and abundant bivalent metal ions in biology. There is therefore great interest in the biological chemistry of these ions; however, having closed-shell electronic structures, they have very few tractable spectroscopic properties such as EPR, UV, or visible spectra, or luminescence with which to probe their solution chemistry. This chapter will deal only with the magnesium ion; however, most of what is said about Mg^{2+} ions applies equally well to Ca^{2+}. There are only a few methods available for direct studies of Mg^{2+} ions, for example, atomic absorption for determination of total magnesium, or ion-selective electrodes for determination of free $Mg^{2+}(aq)$. There are, however, some other indirect methods that will also be discussed in this chapter. The most versatile method for detailed studies of Mg^{2+} ion binding and kinetics is probably ^{25}Mg NMR. This technique has not been widely used, partly because the ^{25}Mg isotope is quadrupolar and has received an undeserved bad reputation as a difficult spin-$\frac{5}{2}$ nucleus. A major part of this chapter will be devoted to this technique with the hope that it will be more frequently used in the future.

   Section 2.2 will briefly review the various techniques that have been used to study magnesium chemistry in a biological context. This is done by selection of specific examples rather than an exhaustive review of the literature. Sections 2.3 and 2.4, which are the major sections of this chapter, will deal with two specific problems: first, how to determine binding constants for Mg^{2+} ion, and second, how to evaluate the kinetics of binding. Finally, I will conclude with a short section on data treatment. One might believe that such a section should not be necessary;

however, in my experience the treatment of raw data might best be described as mistreated.

2.2 Detection Techniques

2.2.1 Atomic Absorption

Flame atomic absorption spectroscopy (FAAS) is very suitable for determination of total magnesium content in biological material. Of course, it is a destructive method, and one needs ~ 1 mL of a solution containing 0.1–4 mg of Mg^{2+}. This corresponds to a concentration between 4 μM and 0.16 mM. Since the Mg^{2+} concentration in biological material typically varies from 1 mM on upwards, the solution requirement will be from a few to 100 μL. One advantage that FAAS affords is that there is normally a minimal requirement for sample preparation. Biological material can in many cases be used as is, but in most instances a dilution (up to 1000-fold) is necessary.

A very detailed description of FAAS can be found in Vogel's textbook,[1] but the basic theory is very simple. Gaseous atoms in a flame will absorb light at frequencies typical of each element (for Mg, $\lambda = 285.2$ nm). A standard acetylene air flame can be used, and there is normally no interference from other elements. The primary output from FAAS is the absorbance, rather than concentration. The conversion of absorbance into concentration can be done in either of two ways.

2.2.1.1 Calibration Curve

A calibration curve for atomic absorption is made with the aid of samples containing well-defined amounts of Mg^{2+} ions in a solution as similar as possible to the test solution. These solutions are aspirated into the flame, and the readings are plotted against the known concentrations. The concentration of Mg^{2+} should preferably be adjusted to obtain absorbances in the range 0.1 to 0.4. If the calibration curve is nonlinear, then more points are needed than for a linear curve. As soon as the calibration curve is constructed, it is a simple matter to convert absorbance readings into concentration. For Mg^{2+}, the test solution often has to be diluted in order to obtain a reading within the calibration range. Nowadays the calibration curve will be in a microcomputer that will output the concentration after the absorbance reading has been entered.

2.2.1.2 Standard Additions

When dealing with test solutions with a very complex composition, it may be difficult to produce a suitable calibration curve. In such cases the test solution is first, if necessary, diluted to obtain an absorbance reading in the lower end of the range, and then additions of well-defined amounts of Mg^{2+} are made stepwise, taking absorbance readings for each step. The magnesium concentration in the test solution can then be obtained from an extrapolation. It must, however, be kept in mind that an extrapolation like this is always less accurate than an interpolation in a calibration curve. This is especially true if the extrapolation has to be done from a curve that is far from linear.

2.2.2 Ion-Selective Electrodes

With an ion-selective electrode, the activity of the magnesium ion [Mg^{2+}] can be measured. This means that Mg^{2+} ions bound to, for example, ATP or proteins will not be detected. During recent years, ion-selective electrodes based on semipermeable membranes have been developed. With such an electrode the activity of a given ion can be measured in aqueous solution. The sensitivies of these electrodes are different for different ions (Mg^{2+} concentrations down to $\sim 10^{-6}$ M can be measured). The measured voltage varies as a function of activity (not concentration) according to the Nernst equation (2.1), where E is the measured total potential, E_o is the portion of the total potential that is due to reference electrodes and internal solutions, R and F are constants with their normal meanings [$(2.3RT)/2F) = 29.6$ mV at 25°C], and A_{Mg} is the activity of the Mg^{2+} ions in the sample.

$$E = E_o + [(2.3RT)/(2F)] \log A_{Mg} \tag{2.1}$$

When accurate determinations are needed, a calibration curve should be produced based on accurately known reference samples. The technique can also be used to obtain rough estimates of concentrations without the use of a calibration curve. For samples of low ionic strength ($< 10^{-4}$ M) the concentration will be almost identical to the activity, and the concentration can be obtained directly. For higher ionic strength the concentration has to be calculated using the activity coefficient, which can be calculated from the Debye–Hückel equation. In an isotonic solution the activity coefficient will be about 0.3, and so there is a significant difference between activity and concentration at higher concentrations. The equipment needed for these experiments is the ion-selective electrode, which can be plugged into a standard pH meter. One has to be aware, however, that these electrodes have a limited lifetime and must be calibrated frequently.

2.2.3 Nuclear Magnetic Resonance

NMR can be used in two different ways in studies on magnesium in biological systems. Magnesium has an NMR-active isotope, ^{25}Mg, with a spin $I = \frac{5}{2}$ that can be used for studies directly on the Mg^{2+} ion.[2] On the other hand, both ^{1}H and ^{13}C NMR can and have been used to study the interaction between Mg^{2+} ion and its ligands. When a Mg^{2+} ion binds to a protein, for example, this binding may cause some structural changes, perhaps very minor ones, that can be detected by ^{1}H or ^{13}C NMR through shift changes. For example, the binding of Mg^{2+} ions to some calcium binding proteins belonging to the calmodulin superfamily have been studied by ^{1}H NMR. Another example where ^{1}H NMR has been used is in studies of body fluids, such as urine. By adding EDTA and then recording an NMR spectrum, the amount of Mg^{2+} and Ca^{2+} can be obtained since these two metal–EDTA complexes give rise to distinct ^1H NMR signals. The theory of NMR is too complex a subject to be dealt with in this chapter; however, there are a multitude of textbooks available covering various levels.[3–5] In this chapter I will only touch upon some of the theory that is of particular interest for ^{25}Mg NMR.

Some of the NMR properties of ^{25}Mg are collected in Table 2.1. Relative to ^1H NMR, the sensitivity is 3×10^{-3}, which is not too bad. The natural abundance also is reasonably high (10%), and ^{25}Mg enriched to 95% is readily available at a reasonable cost. So what is the problem? The main difficulty with ^{25}Mg NMR is that it has a spin quantum number larger than $\frac{1}{2}$ ($I = \frac{5}{2}$). It is therefore a quadrupolar nucleus, and the NMR properties will be dominated by quadrupolar effects. The quadrupole moment may not appear large; however, it is sufficiently large to result in a very efficient relaxation rate whenever the $^{25}Mg^{2+}$ ion is bound to a macromolecule, and consequently will give rise to a very broad NMR resonance that may even be difficult to detect. The relaxation rate of "free" solvated $^{25}Mg^{2+}$ ions in dilute aqueous solution is 4.6 s^{-1}, which may be compared to a rate of 0.5 s^{-1} for $^{43}Ca^{2+}$.[6] There are very few reports on ^{25}Mg NMR studies of Mg^{2+} ion binding to small ligands,[7,8] making it difficult to judge how useful it may be for such studies. Figure 2.1 shows ^{25}Mg NMR spectra of free $^{25}Mg^{2+}$ and $^{25}Mg^{2+}$ bound to EDTA. It is obvious that even for this small complex there is a very strong broadening of the ^{25}Mg resonance. At first glance this may give the impression that NMR is not a suitable method for studies of Mg^{2+} ions in biological systems. This is not true. On the contrary, the strong broadening of the ^{25}Mg NMR signal due to binding of the Mg^{2+} ion to a macromolecule offers a very sensitive probe for studying Mg^{2+}–macromolecule interactions. As will be shown, it can be used in studies of equilibria as well as kinetics.

2.2.3.1 Some Experimental Considerations

The resonance frequency of ^{25}Mg is quite low, 22.1 MHz at 8.45 T (360 MHz for ^1H), resulting in a low sensitivity. High magnetic fields and enrichment are therefore a necessity. Additional improvements in the sensitivity can be obtained by using large sample volumes (20-mm-diameter NMR tubes) and use of sideways-oriented solenoid transmitter–receiver rf coils. Commercial NMR spectrometers with superconducting magnets are routinely equipped with saddle-shaped Helmholtz coils. This coil arrangement is necessary to insert and remove the samples vertically.

Table 2.1   SOME PHYSICAL PROPERTIES OF ^{25}Mg

Spin	5/2
Natural abundance	10.1%
Quadrupolar moment	$0.22 \times 10^{-28} \text{ m}^{-2}$
Resonance frequency at 11.8 T	30.8 MHz
Relative sensitivity (1.0 for ^1H)	3×10^{-3}

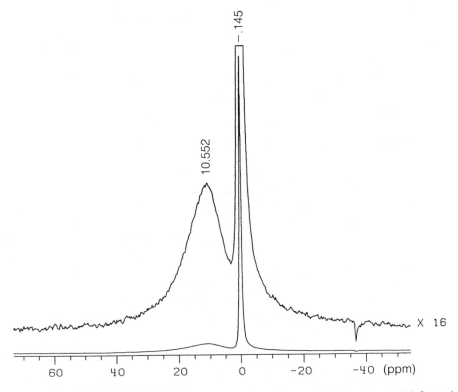

Figure 2.1   A 22-MHz ^{25}Mg NMR spectrum of a mixture containing 10 mM Mg^{2+} and 5 mM EDTA at pH 10 and 40°C.

Although the Helmholtz coils are superior to solenoids in terms of correcting for B_o field inhomogeneity, they are inferior to solenoids when it comes to sensitivity by a factor of ~ 2.[9] Specially built probes with horizontally arranged solenoidal transmitter–receiver coils may now be obtained on special order.

2.2.3.2 *Some Theoretical Considerations*

The NMR relaxation properties of the ^{25}Mg nucleus are completely dominated by the quadrupolar effects. This will in a sense simplify the treatment of the relaxation; however, there may also be some complications. When the correlation time describing the reorientation of the magnesium nucleus, τ_c, is small compared to the inverse resonance frequency, $\omega\tau_c \ll 1$, the so-called extreme narrowing case, the relaxation times T_1 and T_2 are equal, and the relaxation follows a simple exponential curve. We have for this case

$$1/T_1 = R_1 = R_2 = \frac{3\pi^2}{10} \frac{2I + 3}{I^2 (2I - 1)} \chi^2 \left(1 + \frac{\eta^2}{3} \right) \tau_c \qquad (2.2)$$

where $I = \frac{5}{2}$ is the spin quantum number, χ is the quadrupole coupling constant, η is the asymmetry factor, and τ_c is the correlation time. Typically, η lies between 0 and 1, and rarely above 0.5, and so the factor $\eta^2/3$ will be neglected in the rest of this chapter.

The quadrupole coupling constant depends on the electric field gradient at the position of the nucleus and therefore contains information about the local symmetry around the nucleus. In the case of extreme narrowing, only the product $\chi^2 \tau_c$ can be determined directly by ^{25}Mg NMR. In order to determine χ^2, an independent determination of τ_c is required. This may often be obtained from ^{13}C NMR relaxation time measurements. However, one has to bear in mind that the ^{25}Mg nucleus may have some extra freedom of motion not picked up by the ^{13}C nuclei.

When the extreme narrowing condition does not apply, the two relaxation times, T_1 and T_2, are no longer identical. In fact there are no well-defined relaxation times since the relaxation back to equilibrium will be described by a multiexponential shape. Bull and co-workers have shown that, in practice, for nuclei with $I > 1$ relaxation often appears exponential, even though, from Eqs. (2.3) and (2.4) the measured longitudinal and transverse relaxation times are different.[10,11] Useful expressions for the apparent T_1 and T_2 can be derived by linearization and are valid for $\omega\tau_c \leq 1.5$, where $J_n = \tau_c/[1 + (n\omega\tau_c)^2)]$. Similarly, Halle and Wennerström have shown for $I = \frac{5}{2}$ that there is an approximately exponential relaxationship up to $\omega\tau_c = 1$.[12]

$$R_1 = \frac{3\pi^2}{10} \chi^2 \frac{2I + 3}{I^2(2I - 1)} (0.2J_1 + 0.8J_2) \tag{2.3}$$

$$R_2 = \frac{3\pi^2}{10} \chi^2 \frac{2I + 3}{I^2(2I - 1)} (0.3J_0 + 0.5J_1 + 0.2J_2) \tag{2.4}$$

Also, for systems where a large excess of free solvated ions are found in fast exchange with metal ions bound to sites with long correlation times, the observed resonances will relax with an apparent relaxation rate that is exponential, even though the relaxation of the ions bound to the macromolecule may be far from exponential.[11] When both R_1 and R_2 can be measured, it is thus possible to determine both χ and τ_c, and in this way it is possible to obtain some information regarding the symmetry of the metal binding site. This has been done in quite a few cases for ^{43}Ca, and it could in principle be applied to ^{25}Mg.[13] ^{25}Mg NMR has been used, as will be discussed, to determine both binding constants and exchange rates. Another complication that can appear in spectra for nuclei with a quadrupole moment is that the shift may become field dependent.[14] This appears only outside the extreme narrowing regime and is called the second-order dynamic frequency shift. To my knowledge this has not yet been observed for ^{25}Mg.

2.2.4 Magnesium Chelators

There exist metal ion chelators based on EDTA and EGTA that have spectroscopic properties suitable for determination of metal ion concentration. QUIN 2 (2-{[2-bis (carboxymethyl)-amino-5-methylphenyl]methyl}-6-methoxy-8-bis(carboxymethyl)-amino quinoline) has a fluorescence spectrum that changes upon metal ion binding.[15] BABTA (1,2-bis(o-aminophenoxy)-ethane-N,N,N',N'-tetraacetic acid) and substituted BABTAs have absorption spectra that are dependent on metal ion binding. They can both be used to measure metal ion concentration. At a suitable wavelength we will have the following equations:

$$A^{obs} = (p^f A^f + p^b A^b) [Ch]_{tot} \tag{2.5}$$

$$[Ch]^b = p^b [Ch]_{tot} = (A^{obs} - A^f) [Ch]_{tot}/(A^b - A^f) \tag{2.6}$$

where A^{obs}, A^f, and A^b are the observed absorbance and the absorbance for free and bound chelator, respectively; p^f and p^b are the population of free and bound chelator, respectively; and $[Ch]_{tot}$ is the total chelator concentration. Both A^f and A^b can be determined in separate experiments with either metal-free chelator or chelator saturated with metal ions.

If the binding is sufficiently strong to result in quantitative binding of Mg^{2+} ions, the Mg^{2+} ion concentration will be given by

$$[Mg^{2+}] = [Ch]_{tot} p^b \tag{2.7}$$

If, on the other hand, the binding is not so strong, then the Mg^{2+} concentration is given by Eqs. (2.8) or (2.9). None of the above chelators is selective for Mg^{2+} over Ca^{2+}; so if there are comparable amounts of Ca^{2+} and Mg^{2+} present in solution, then this method may not be used. However, if the Mg^{2+} concentration is much higher than the Ca^{2+} concentration, as is typically the case inside cells, the effect due to the Ca^{2+}–chelator complex can be corrected for, if necessary.

$$[Mg^{2+}] = p^b/(p^f K) \tag{2.8}$$

$$[Mg^{2+}]_{tot} = p^b(1 + p^f K[Ch_{tot}])/p^f K \tag{2.9}$$

Equation (2.8) is useful when we know that the chelator has a much higher affinity for Mg^{2+} ions than for any other ligand present in the solution. This is of course not known beforehand. To check if this is correct, one should perform a set of experiments with increasing chelator concentrations. If the determined Mg^{2+} concentration levels off at higher chelator concentrations, then this is a good indication that the correct value has been found. It is, however, not absolutely certain since there may still be some magnesium ion bound even more strongly to some other ligand that will not appear in these measurements.

On the other hand, if one uses a very weak chelator to determine free Mg^{2+}, one has to be very careful not to use so much of the chelator that the equilibrium in the system is seriously altered. There are examples in the literature where this technique has been used in a very uncritical way. Whether one uses a strong chelator to determine total Mg^{2+}, or a weak chelator to determine free Mg^{2+}, controls must be performed to ensure the validity of the results. As will be discussed further, chelators are also useful in detailed studies of Mg^{2+} ion binding to proteins or other ligands as well as a probe in stopped-flow studies.

2.3 Equilibria

In studies of magnesium biochemistry, the types of equilibria normally considered include metal ion binding to ligands such as proteins, and ADP or ATP. In order to understand the function of the Mg^{2+} ion in these systems, it is of course necessary to know the relevant binding constants. In the following I will discuss some of the techniques that have been, or can be used to this end.

2.3.1 Equilibrium Dialysis

Equilibrium dialysis appears to be a popular method among biochemists to study various equilibria between a small ligand, for example, a metal ion, and a macro-molecule. The macromolecule–ligand solution is placed in a dialysis bag that al-lows free passage of the small ligand, but not the macromolecule. This bag is placed in a bath containing suitable buffer solutions, and the system is allowed to reach its equilibrium point. At this stage we need a suitable method to analyze the total Mg^{2+} concentration both inside and outside the bag. At equilibrium the concentration of free Mg^{2+} will be the same in the two solutions, or more accurately, the chemical potential of Mg^{2+} will be the same in the two compartments. The excess Mg^{2+} inside the dialysis bag is the result of Mg^{2+} ions bound to the macromolecule. By performing a series of such experiments, using different total concentrations of Mg^{2+} in the starting solutions, the binding constant may be calculated so long as it is in a suitable range as compared to the total concentration of Mg^{2+} and the macromolecule. Ideally the concentration of the macromolecule should be similar to the inverse of the binding constant, $1/K_B = K_D$ (the dissociation constant). In most cases the Mg^{2+} binding constant is not very high ($K_B < 10^6$ M^{-1}), and the Mg analysis can be made with FAAS.

Plotting the fraction of saturated binding site(s) as a function of free magnesium results in a sigmoidal curve (Fig. 2.2) with a midpoint corresponding to K_D. This is a familiar titration curve; however, it is often not the best way to treat the data since for high concentrations the fraction of saturated binding site(s) will be obtained from a small difference between two large numbers. The accuracy in the determined fractional saturation varies, therefore, dramatically with the total concentration of Mg^{2+}. A better way is to plot the two determined concentrations directly, one versus the other (Fig. 2.3), and calculate a theoretical curve that fits the experimen-tal data. In this way the data have not been manipulated, and there is no false weighting of the data. There is of course nothing wrong in drawing the traditional titration curve that we are used to seeing. It is, however, necessary to keep in mind that the errors in the experimental data points for the higher concentrations may be very large. Often the accuracy is limited by the determination of protein concentra-tion. The problem here is clearly seen from reported noninteger values for the number of binding sites in "pure" proteins.

2.3.2 Titration with Fluorescent Chelators

As mentioned previously, there are chelators based on the EDTA and EGTA struc-tures that are fluorescent due to the insertion of aromatic rings in the molecules. The

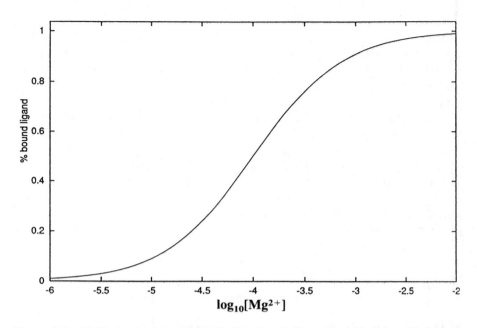

Figure 2.2 Mg^{2+} titration of a Mg^{2+}-binding ligand. $K_D = 10^{-4}$ M, [L] = 10^{-4} M. A sigmoidal titration curve where the percent saturation of the ligand is plotted versus the logarithm of the free Mg^{2+} concentration.

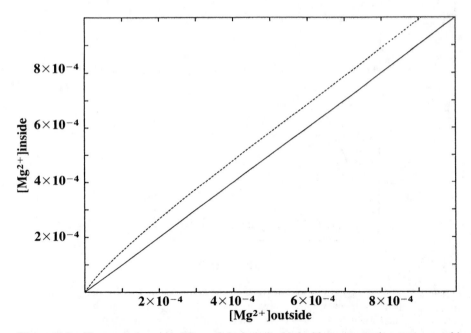

Figure 2.3   The same data as in Figure 2.2 plotted with inside concentration versus outside concentration. The solid line is a reference line with no Mg^{2+} ligand inside. The difference between the solid and dashed lines is bound Mg^{2+}, as plotted in Figure 2.2.

fluorescence spectra of these chelators changes upon metal binding. A very accurate method to determine binding constants for metal ions to proteins has been developed.[16] The method was developed for Ca^{2+} but applies as well for Mg^{2+} under suitable conditions. "Suitable conditions" means in this case that one has to find a chelator that has its Mg^{2+} binding constant within a factor of 10 from that of the protein.

In a solution containing Mg^{2+} ions, a chelator, and a magnesium binding protein, the Mg^{2+} ions will be distributed between the chelator and the protein in a manner that depends on their relative binding constant. Subsequently, the amount bound to the chelator can be obtained from the fluorescence spectrum. The only instrumentation needed for this experiment, therefore, is a suitable spectrophotometer. Schematic titration curves are shown in Figure 2.4. The upper curve is for the chelator alone with no protein present, and the lower curve is for a system containing equal numbers of binding sites in the chelator and protein, with the same binding constant to both chelator and protein. Before the metal ion binding constant to the protein can be determined, the binding constant for the chelator has to be

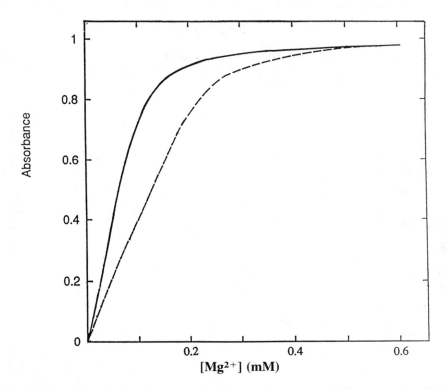

Figure 2.4 Schematic titration curves showing absorbance versus Mg^2 concentration. Solid curve: only the fluorescent chelator is present. Dashed curve: chelator plus the Mg^{2+} binding protein. For the dashed curve the chelator and the protein are assumed to have both the same Mg^{2+} affinity and an equal number of binding sites. The absorbance change is assumed to be exclusively caused by Mg^{2+} binding to the chelator.

known. Linse et al. have shown that for Ca^{2+} the chelator's affinity for metal ions depends strongly on the ionic strength of the solution.[16] It decreases by a factor of about 20 when changing from low salt to 0.15 M KCl, when all calculations are made on a concentration basis. It is therefore important to determine the binding constant for the chelator at the appropriate ionic strength. In the case of Mg^{2+}, where the affinity of the chelator is not too high, this can be done directly in a Mg^{2+} titration using a chelator concentration similar to the expected dissociation constant. The binding constant is then best obtained from a least-squares fit of calculated data points to the experimental ones. As can be seen in Figure 2.5, binding constants that are up to 10 times stronger than the inverse of the chelator concentration can be determined in this way. This chelator can then be used to determine Mg^{2+} binding to, for example, a protein. Mg^{2+} is titrated into a solution containing comparable amounts (on a molar basis) of protein and chelator.

Since magnesium binding is competitive between the ligand and the chelator, the distribution of Mg^{2+} ions between its various binding sites will be determined by the relative binding constants. Some theoretical titration curves are shown in Figure 2.6. The Mg^{2+} binding constant to the protein can now be obtained in a least-

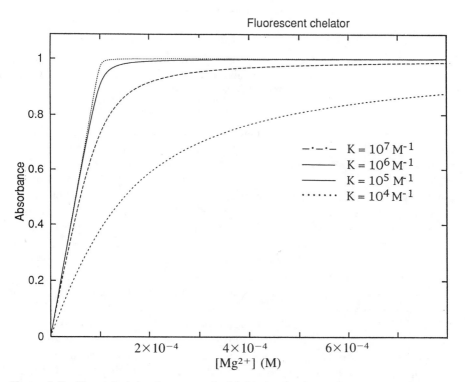

Figure 2.5 Theoretical titration curves for Mg^{2+} titration into a fluorescent chelator. The chelator concentration is 10^{-4} M, and the Mg^{2+} binding constant varies from 10^4 to 10^7 M^{-1}.

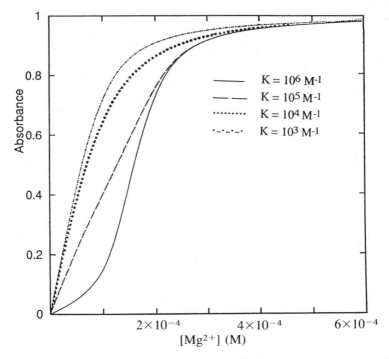

Figure 2.6 Theoretical titration curves showing absorbance versus Mg^{2+} concentration with assumed chelator and protein concentrations of 10^{-4} M. The binding constant for the chelator was 10^5 M^{-1}, and for the protein it has been varied from 10^3 to 10^6 M^{-1}.

squares fit of calculated to experimental absorbances as a function of total Mg^{2+} concentration. Of course this calculation is model dependent since the number of binding sites in the protein must be assumed. The normal procedure is to start with the simplest possible model in agreement with other knowledge. If this model will not result in an acceptable agreement between calculated and experimental data, a more complex model has to be assumed. One will in the end accept the most simple model that will satisfactorily explain the experimental data. One prerequisite for this treatment is that the protein concentration is well known. In my experience this is not normally the case, and if there are more than four binding sites in the protein, the number will not be determined reliably.

The details in the calculations can naturally vary; however, I generally prefer the following one based on the use of Eqs. (2.10)–(2.12), which are valid for a protein with two binding sites. For each data point the concentrations of all species in the solution have to be calculated from known or assumed total concentrations and binding constants. For all but the most simple case this has to be done in an iterative fashion. Starting with assumed values for the free concentrations of [Mg], [Ch], and [L], and knowing their total concentrations, we can iteratively solve Eqs. (2.10) to (2.12) until self-consistency is reached. This will give [Mg·Ch] ($= K^{Ch}$[Mg][Ch]), which is proportional to $A - A_o$, which is our observable.

$$[Mg] = Mg^{tot}/(1 + K^{Ch}[Ch] + K_1[L] + K_1K_2[L][Mg]) \tag{2.10}$$

$$[Ch] = Ch^{tot}/(1 + K^{Ch}[Mg]) \tag{2.11}$$

$$[L] = L^{tot}/(1 + K_1[Mg] + K_1K_2[Mg]^2) \tag{2.12}$$

The next step is to calculate the error square sum.

$$ESS = \sum_i (Abs_{o,i} - Abs_{c,i})^2 \tag{2.13}$$

The unknown binding constant(s) are now varied until the minimum in ESS is found. This can be done with any standard procedure. With the currently common availability of inexpensive computers, there is no need to try to find linearized equations, with the unavoidable modification of the raw data, instead of fitting the raw data directly as suggested.

2.3.3 NMR

^{25}Mg NMR can be used to study the binding of Mg^{2+} ions to macromolecules. The ^{25}Mg chemical shift changes observed to date are quite small and may not be useful (see Fig. 2.1); however, the relaxation rates vary significantly upon binding to macromolecules. The width of the ^{25}Mg resonance from free Mg^{2+} ions is rather narrow, whereas that from Mg^{2+} ions bound to a macromolecule is very broad. The width of the observed signal will therefore be very sensitive to the amount of bound Mg^{2+}. In the case of fast exchange between free and bound ions the observed linewidth is given by

$$\Delta\nu^{obs} = p^b \Delta\nu^b + p^f \Delta\nu^f \tag{2.14}$$

where $\Delta\nu^{obs}$, $\Delta\nu^b$, and $\Delta\nu^f$ are the observed linewidth and the linewidth for bound and free ions, respectively, and p^b and p^f are the populations of bound and free ions, respectively. Figure 2.7 shows some examples of theoretical Mg^{2+} titration curves.

Since the sensitivity of NMR results in a lower detection limit of approximately 0.1 mM for ^{25}Mg, we are also limited to determining binding constants of less than 10^5 M^{-1}. For such strong binding the exchange between free and bound Mg^{2+} ions is normally not fast enough for Eq. (2.14) to be applicable. In the other extreme, when the exchange is very slow, two NMR signals may be observed during the titration: one narrow signal from free ions and a broad signal from bound ions. In such a case the intensity of the signal from free ions may be used to calculate the binding constant. The signal from bound ions is often too broad even to be detected. Figure 2.8 shows spectra obtained from a titration of Mg^{2+} ions into a solution containing a calcium binding protein (the C-terminal half of troponin C). Such results are not easily used to extract reliable binding constants; however, the results agreed generally with other measurements.

More complex equilibria can also be studied by ^{25}Mg NMR. Thus Wahlgren and Drakenberg have used ^{25}Mg NMR to study the binding of Mg^{2+} to β-casein.[17] Various other techniques have previously been used to study metal ion binding to the

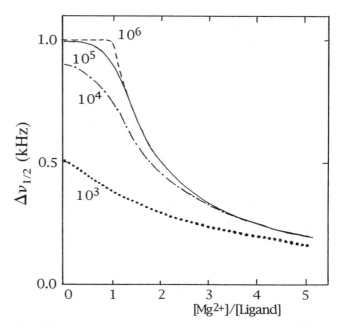

Figure 2.7 Theoretical titration curves showing the dependence of the ^{25}Mg NMR line-width on the Mg^{2+} concentration for various binding constants, as indicated. The ligand concentration is assumed to be 10^{-3} M. Binding constants up to 10^5 M^{-1} can be determined in this way if the exchange is fast, $k_{off} > 10^3$ s^{-1}, as assumed in the calculations.

casein proteins. β-Casein has a cluster of four phospho-serines that are supposed to be mainly responsible for the metal binding properties of this protein. The results from a Mg^{2+} titration to a β-casein solution, detected by ^{25}Mg NMR, is shown in Figure 2.9. The experimental data clearly show that binding occurs in two distinct steps. Simulation showed that the experimental data could be fitted with a model of one strong and four weaker binding sites. Furthermore, the calculation showed that the strongly bound Mg^{2+} ion is not in fast exchange with free ions. Equation (2.14) is therefore not applicable, and a model taking exchange into account had to be used (see a following discussion of NMR).

Both ^1H and ^{13}C NMR can also be used to study Mg^{2+} binding. The Lund group has used ^1H NMR to study the binding of Mg^{2+} to various calcium binding pro-teins.[18] It was possible to show in some of these studies that Mg^{2+} binding to the two sites in the C-terminal half of troponin C is sequential, whereas Ca^{2+} binding has been found to be strongly cooperative. Furthermore, Cowan and co-workers have used ^1H NMR to study Mg^{2+} binding in various systems.[19]

2.3.4 Thermodynamic Measurements

All chemical and biochemical reactions can in principle be followed by thermo-dynamic measurements. Until recently, the sensitivity of microcalorimeters set a

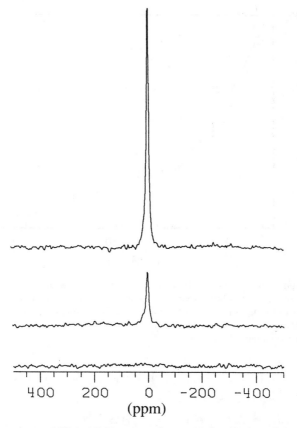

Figure 2.8 Experimental ^{25}Mg NMR spectra from a Mg^{2+} titration of the C-terminal half of the calcium-binding protein troponin C. Note that the signal from ^{25}Mg bound to the protein is too broad to be observed. The sharp signal shown is from free $^{25}Mg^{2+}$ ions in slow exchange with the bound ones.

limit to what binding constants could be reliably determined to be less than 10^5 M^{-1}. Recent developments in the area have now made it possible to determine binding constants as high as 10^8 M^{-1}. This is certainly high enough for most, if not all, metal ion binding involving Mg^{2+} ions.

In principle the thermodynamic measurements are very simple. In a titration of, for example, Mg^{2+} into a solution containing a Mg^{2+}-binding protein, the released heat is measured for each point in the titration. The tricky part of the experiment is the determination of the amount of released heat. In the newly developed computerized titration calorimeter by MicroCal (Northamton, Massachusetts), heat released down to the order of 0.1 mcal can be measured.[20]

For the case of a single binding site, data analysis is straightforward. After making the necessary corrections for the heat of mixing, the released heat is normally plotted as a function of time, or titration point number, which of course can

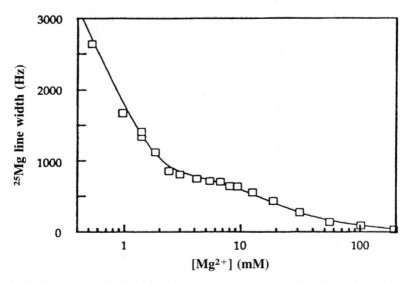

Figure 2.9 Experimental ^{25}Mg NMR data showing the linewidth of the observed ^{25}Mg signal in the presence of the milk protein β-casein. The solid curve is a best-fit calculated curve with the assumption of two classes of sites. The first class contains one site with strong binding and intermediate exchange, whereas the second class contains four sites with weaker binding and fast exchange.

easily be converted into concentration. The experimental curve can now be fitted with a theoretical curve given by

$$\Delta\Delta H_n = \Delta H_n - \Delta H_{n-1} = (p_n - p_{n-1}) \Delta H° \tag{2.15}$$

where $\Delta\Delta H_n$ is the heat released after the addition number n, and ΔH_n and ΔH_{n-1} are the integrated heat released up to and including titration point number n and $n-1$, respectively, p_n and p_{n-1} are the proportions of occupied binding sites after titration point number n and $n-1$, respectively, and $\Delta H°$ is the total amount of released heat when the binding site is saturated (Fig. 2.10).

Moeschler et al. have used thermodynamic measurements in their studies of Ca^{2+} and Mg^{2+} binding to parvalbumin.[21] By combining the enthalpy of reaction obtained in the thermodynamic study with the free energy calculated from the binding constants ($\Delta G = RT \ln K$), the entropy of the reaction could be calculated from $\Delta S = (\Delta H - \Delta G)/T$. In this study the binding constant for Ca^{2+} ions to parvalbumin was too high to be obtained from microcalorimetric data and were therefore taken from an equilibrium dialysis experiment. The magnesium binding constant was obtained from a competition experiment. Using the modern titration calorimeter described, it would have been possible first to determine the Mg^{2+} binding constant directly by titrating Mg^{2+} into a protein solution. The calcium binding could then have been obtained by titrating Ca^{2+} into a solution containing protein and Mg^{2+} in excess. The expected titration curves are shown in Figure 2.11.

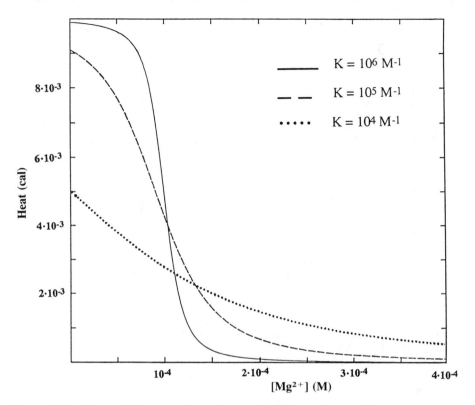

Figure 2.10 Simulated isotherms for a microcalorimetric experiment. The assumed concentration of Mg^{2+}-binding sites is 10^{-4} M. Curves are shown for different binding constants as indicated.

2.4 Kinetics

Kinetic measurements must be considered in discrete groups according to how fast the exchange occurs. In the particular case of studies with Mg^{2+} ions, one only observes exchanges where the ions move between distinct environments, such as between free solution or protein bound. Most likely it will not be possible to vary the enzyme concentration to moderate the formation of reaction products, as is common in enzyme kinetic studies. For example, the off-rate of Mg^{2+} ions from its binding site(s) in a protein will normally depend only on the temperature, and can therefore be varied only within a very limited range.

Very slow reactions ($k_{off} < 10^{-2}$ s^{-1}) may be studied by any method that can detect free Mg^{2+}, and so follow the change in that concentration, for example, an ion-selective electrode. If the exchange is extremely slow, then it may be possible to isolate the free Mg^{2+} by dialysis, and thereafter any method that detects either free Mg^{2+} or total Mg can be used. For faster reactions, $k_{off} < 100$–1000 s^{-1}, the

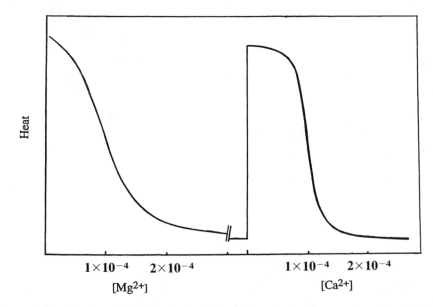

Figure 2.11 Simulated isotherm for a microcalorimetric experiment where first Mg^{2+} is titrated into a 5×10^{-5} M parvalbumin solution, followed by a Ca^{2+} titration. Assumed binding constants are $K^{Mg} = 10^5$ M^{-1} and $K^{Ca} = 3 \times 10^9$ M^{-1}. A competitive binding model and no cooperativity are assumed.

standard technique is stopped flow with a suitable detection method. When the reaction is too fast for stopped flow, ^{25}Mg NMR can be used. This technique differs from the aforementioned examples in that it measures the rate(s) under equilibrium conditions. Exchange rates ranging from 10 to 10^5 s^{-1} can be measured in this way.

2.4.1 Slow Kinetics

Most likely there will be only a few, if any, cases where the Mg^{2+} exchange is slow enough for use of techniques where samples can be taken out of the reaction vessel to be analyzed. Accordingly, this will not be discussed further here.

2.4.2 Stopped Flow

In stopped-flow kinetic studies two or more liquids are mixed quickly, and a property that changes during the reaction is measured. To have sufficiently fast detection there must normally be a change in the spectroscopic properties of one of the reactants that can be monitored, such as the UV absorption of a protein upon binding of a metal ion due to a conformational perturbation of the environment of an aromatic side chain or a label. Another possibility is that a released species (for example, Mg^{2+} ion) binds to a fluorescent chelator (see the earlier section on magnesium chelators). The limiting factor in these studies is the speed with which the mixing can be performed. In a modern stopped-flow apparatus the mixing may be complete within one millisecond. This means that a kinetic process with a half-life of at least 1 ms can be studied.

Bayley and co-workers have used this technique to study the kinetics of metal binding to calcium binding proteins belonging to the calmodulin family of proteins.[22–25] When one solution containing the Ca^{2+} (or Mg^{2+}) loaded protein is mixed with a solution containing a fluorescent chelator, the metal ion off-rate(s) can be determined from the time dependence of the fluorescence, as shown in Figure 2.12. In order to evaluate such curves, it is first necessary to confirm that the metal binding to the chelator does not contain any rate-limiting, slow steps. This is most conveniently done in a control experiment where the metal ion containing solution does not contain any protein. In this experiment the reaction should preferably be complete within the mixing time of the stopped-flow apparatus, or at least be much faster than the reaction in the presence of protein. Second, the results are most easily interpreted if there is no detectable back reaction, that is, if the chelator is a much better metal binder than the protein. Under such favorable conditions, the metal ion off-rate from the protein (k_{off}) is given by the following equation and can be obtained by fitting this equation to the experimental data.

$$A^{obs} = (A - A_o)(1 - e^{-tk_{off}}) + A_o \qquad (2.16)$$

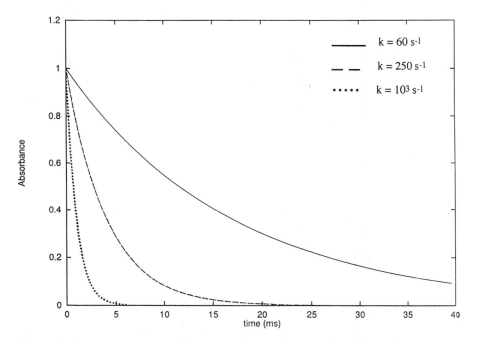

Figure 2.12 Simulated stopped-flow experiment where the absorbance of a chelator is followed after mixing the chelator with a Mg^{2+}-loaded protein. Rate constants, k_{off}, are as indicated.

Equation (2.16) can be extended to the case where there are more than one metal-ion binding sites in the protein and the off-rates are different. It is, however, well known that two or more exponential functions with only slightly different time constants cannot be easily resolved. A difference of at least a factor of 2–3 is necessary for a reliable determination. Figure 2.13 shows what the stopped-flow response curve will look like in the case of two different off-rates differing by a factor of 10. The standard procedure in such cases is to use two different time scales on the x axis, as in Figure 2.13. Figure 2.14 shows a theoretical case where there are two metal ions with off-rates of 4×10^2 and 2×10^2 s^{-1}, respectively. The same diagram shows the calculated curve using the same off-rates for the two ions, 2.8×10^2 s^{-1}. As can be seen from this diagram, it is, in fact, not possible to resolve these two rate constants that differ by only a factor of 2.

2.4.3 NMR

NMR has been used extensively to study dynamic processes,[26] but differs from most other techniques used in such studies insofar as the exchange is monitored under

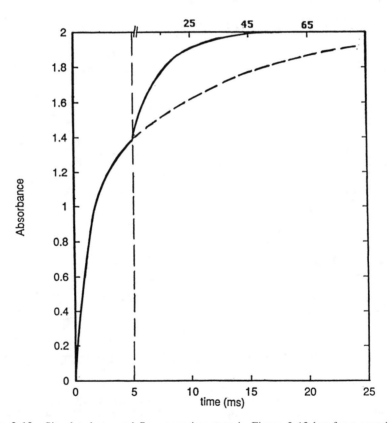

Figure 2.13 Simulated stopped-flow experiment as in Figure 2.12 but for a protein with two Mg^{2+} sites with off-rates of 10^3 and 10^2 s^{-1}.

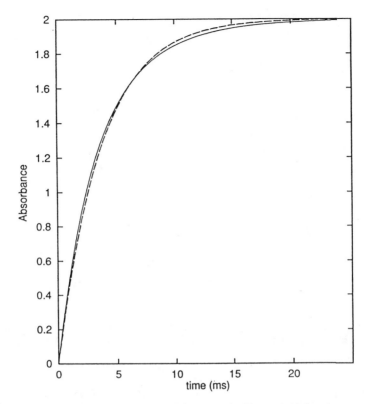

Figure 2.14   Simulated stopped-flow experiment as in Figure 2.12 but for a protein with two Mg^{2+} sites with off-rates of 4×10^2 and 2×10^2 s^{-1} (solid curve). The dashed curve shows the simulated curve for two sites with the same off-rate, 2.8×10^2 s^{-1}.

equilibrium conditions. Every species in a solution will in principle give rise to its own NMR resonance, and if they are sufficiently different from all other resonances, then they will be resolved. However, if the species is involved in a chemical exchange process, the shapes of the resonances will be modified. For the sake of simplicity we will consider here a simple exchange between two states, for example, a metal ion exchanging between free solution or bound to a ligand. When the exchange rate is very slow ($k_{ex} < 1$ s^{-1}) we observe separate resonances for the two environments. Increasing exchange rates will first result in broadening of the resonances, $\Delta\Delta\nu_{1/2} = k_{ex}/\pi$, and as long as $\Delta\nu_{1/2}$ is less than the shift difference between the resonances, we will observe two separate resonances. At still faster exchange rates the two resonances coalesce into one broad resonance that will ultimately become sharper. Very fast exchange rates result in a single Lorentzian resonance with shifts and widths that are the weighted averages of the two resonances (Fig. 2.15).

In the particular case of quadrupolar nuclei such as ^{25}Mg, the main difference between the various resonances are not the chemical shifts but rather the relaxation rates (linewidth). All chemical exchange studies involving Mg^{2+} ion have been

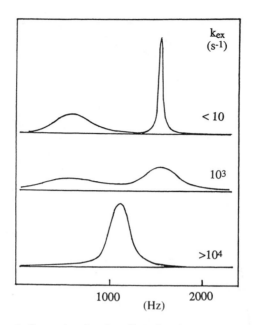

Figure 2.15   Schematic figure showing the effect of exchange on an NMR spectrum consisting of two resonances, such as from free Mg^{2+} and Mg^{2+} bound to a ligand.

exchanges between free ions and ions bound to biological macromolecules. In such cases there is an exchange between a sharp resonances from free ions and a very broad one from bound ions. There is no observable chemical shift difference between the two signals and the broad one is normally too broad to be observed. The band shape for such a case can be derived from the Bloch equations modified for exchange.

$$G_A[1/T_{2A} + k_{AB} - i(\omega - \omega_A)] = ip_A + G_B k_{BA} \qquad (2.17)$$

$$G_B[1/T_{2B} + k_{BA} - i(\omega - \omega_B)] = ip_B + G_A k_{AB} \qquad (2.18)$$

These equations can easily be solved for $G_A + G_B$; however, for more complex exchanges among more than two sites this becomes quite cumbersome, and it is more convenient to let the computer solve the problem directly with complex arithmetical algorithms.

In the case of slow exchange we will only observe the resonance from free ions (Fig. 2.8). With increasing exchange rates this resonance will broaden due to exchange with the bound ions, until the width levels off at the weighted average of bound and free ions. In most if not all kinetic studies involving Mg^{2+} ions, a change in exchange rates will be brought about by a change in temperature. Since the width of the resonances, according to Eq. (2.4), depends on the correlation time (τ_c), which is temperature dependent, a typical temperature dependence will be like the

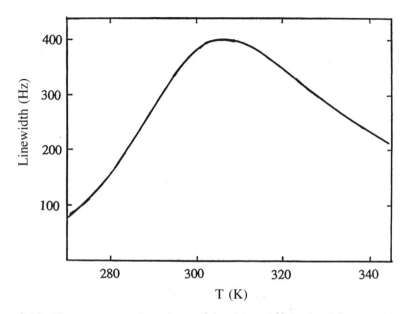

Figure 2.16   The temperature dependence of the observed ^{25}Mg signal for a system where there is an excess of free Mg^{2+} and the resonance from bound Mg^{2+} is too broad to be observed. At low temperatures the observed signal comes from free Mg^{2+} ions and is broadened due to exchange. At high temperatures the observed signal is the average of the resonances from free and bound ions, and the decreasing linewidth is caused by the temperature dependence in the width of the "bound" signal.

one shown in Figure 2.16. At the low-temperature side of the maximum, the observed resonance is mainly that of free ions broadened due to exchange with bound ions, whereas on the high-temperature side of the maximum, both free and bound ions contribute to the observed resonances and the width will be given by Eq. (2.19). The linewidth $\Delta\nu^{b}$ is temperature dependent as a result of the temperature dependence in τ_{c}, which is assumed to be defined by a transition-state type of Eq. (2.20).

$$\Delta\nu^{obs} = p^{b}\,\Delta\nu^{b} + p^{f}\,\Delta\nu^{f} \tag{2.19}$$

$$1/\tau_{c} = (kT/h)\,\exp(-\Delta G/RT) \tag{2.20}$$

The magnitude of τ_{c} may be obtained at a specified temperature from NMR, or some other method, or estimated using the Debye–Stokes–Einstein equation:

$$\tau_{rot} = 4\pi h R_{M}^{3}/3kT \tag{2.21}$$

^{25}Mg NMR has been successfully applied to study Mg^{2+} binding to calmodulin and its tryptic fragments.[27] In this study it proved possible to determine both binding constants and exchange rates for the two classes of sites, as can be seen from Figure 2.17.

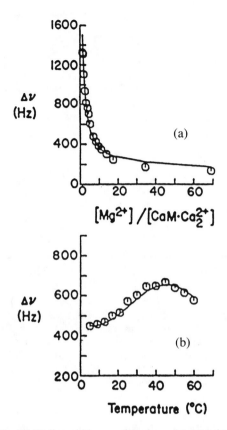

Figure 2.17 The ^{25}Mg NMR linewidth as a function of (a) Mg^{2+} concentration and (b) temperature. In (a) the $[CaM \cdot Ca_2^{2+}]$ concentration varied from 0.98 to 0.78 mM. In (b) the ratio $[Mg^{2+}]/[CaM \cdot Ca_2^{2+}]) = 5.2$. Circles are experimental data points, and the solid curves are the best-fit calculated curves when data from (a) and (b) are treated simultaneously.

2.5 Data Treatment

A rule of thumb when evaluating binding constants or rate constants is to use raw data. Do not try to linearize your data! Nowadays there are computer programs available for PCs or MACs that carry out these calculations, and so there is no longer any need to try to find linear relationships.

Consider an equation of the form $y = 1/(ax + b)$, which is valid for a bimolecular chemical reaction. This may be, and often is, linearized to the form $1/y = ax + b$. In Figure 2.18 I have plotted these two equations, including error bars, with an assumed constant uncertainty in the measurement of y that is 5% of the initial value. As can be seen from this diagram, the error bars for the linearized plot do not extend equally far on the two sides of the line. This will in itself, even for completely random errors, create a systematic error in the calculated slope. The case shown

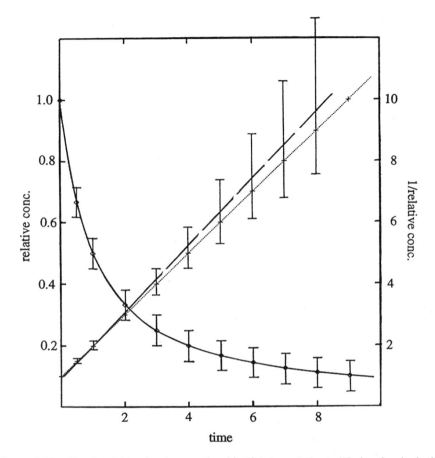

Figure 2.18 Simulated bimolecular reaction kinetics $[y = 1/(x + 1)]$ showing both the direct plot of concentration of reactant versus time (solid line) and the linearized plot with inverse concentration versus time (dotted line). The dashed line shows the expected result with a completely random error of 5% of the initial concentration.

will, if the reaction is followed until 90% completion, result in an overestimation of the slope by approximately 10%. A similar effect is observed in semilogarithmic plots.

More serious errors can be introduced when there are systematic errors in the data. Using the same data shown in Figure 2.18, but now also introducing a 5% systematic overestimation of the measured concentration, results in Figure 2.19. By following the reaction to 80% completion, an underestimation of the slope by 25% is made. By following the reaction to near completion and plotting the data without any manipulations, the systematic error is easily detected and corrected for. One might argue that the same can be done using the linearized form of the equation; however, to correct for the systematic error a computer analysis still has to be done and the "advantage" with the linear equation is completely lost.

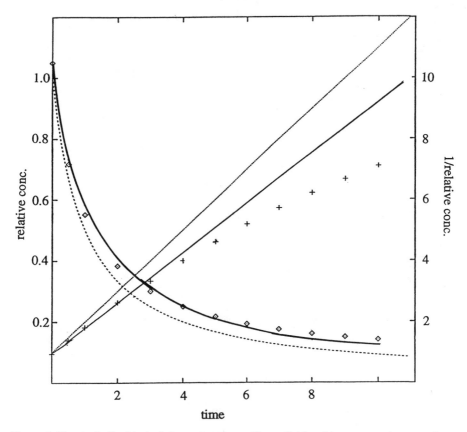

Figure 2.19   A similar kind of plot to that shown Figure 2.18, with a systematic overestimation of the concentration of 5% of the initial concentration, $y = 1/(x + 1) + 0.5$. Dotted curves are those expected from $y = 1/(x + 1)$, and the solid line for the direct plot is for the best fit to $y = 1/(ax+b)$ using all data points and $a = 0.75$ and b 0.95. The solid line for the linearized plot uses only the first four points, resulting in a slope of 0.85. Using all data points gave a slope of 0.6.

The preceding summarizes only two examples of the dangers that may be found with the linearized plots. Since there are now computer programs readily available to do analyses of nonlinear plots, there is really no need for linearization!

References

1. Vogel's Textbook of Quantitative Chemical Analysis, (Eds.) Jeffrey, G. H.; Bassett, J.; Mendham, J.; Demey, R. C., Longman Scientifique & Technical, 1989, pp. 779–813.

2. Drakenberg, T.; Forsén, S., in *A Multinuclear Approach to NMR Spectroscopy,* Lambert, J. B.; Riddell, F. G. (Eds.), D. Reidel Publishing Co., Boston, 1983, pp. 309–28.

3. Derome, A. E., *Modern NMR Techniques for Chemical Research,* Pergamon Press, Oxford, 1990.

4. Harris, R. K., *Nuclear Magnetic Resonance Spectroscopy*, Longman Scientific & Technical, Avon, 1988.

5. Goldman, H., *Quantum Description of High-Resolution NMR in Liquids*, Oxford Science Publications, Oxford, 1988.

6. Lndman, B.; Forsen, S.; Lilja, H., *Chem. Scr.* **1977**, *11*, 91.

7. Robertson, J. P., Jr.; Hiskey, R. G.; Koehler, J., *J. Biol. Chem.* **1978**, *253*, 840–50.

8. Rose, D. M.; Bleam, M. L.; Record, M. T., Jr.; Bryant, R. G., *Proc. Natl. Acad. Sci. (USA)* **1980**, *77*, 6289–95.

9. Hoult, D., *Prog. NMR Spectrosc.* **1978**, *12*, 45.

10. Bull, T. E., *J. Magn. Resonance* **1972**, *8*, 344.

11. Bull, T. E.; Forsén, S.; Turner, D., *J. Chem. Phys.* **1979**, *70*, 3106–11.

12. Halle, B.; Wennerström, H., *J. Magn. Resonance* **1981**, *44*, 89.

13. Andersson, T; Drakenberg, T.; Forsén, S.; Thulin, E.; Swärd, M., *J. Am. Chem. Soc.* **1982**, *104*, 576–80.

14. Arammi, J. M.; Vogel, H. J., *J. Am. Chem. Soc.* **1993**, *115*, 245–52.

15. Tsien, R., *Biochemistry* **1980**, *19*, 2396–404.

16. Linse, S.; Johansson, C.; Brodin, P.; Grundström, T.; Drakenberg, T.; Forsén, S., *Biochemistry* **1991**, *30*, 6723–35.

17. Wahlgren, N. M.; Dejmek, P.; Drakenberg, T., *J. Dairy Res.* **1993**, *60*, 65.

18. Drakenberg, T.; Forsén, S.; Thulin, E.; Vogel, H. J., *J. Biol. Chem.* **1987**, *262*, 672–78.

19. (a) Cowan, J. A., *J. Am. Chem. Soc.* **1991**, *113*, 675–76. (b) Huang, H-W.; Cowan, J. A., *Eur. J. Biochem.* **1994**, *219*, 253–260. (c) Reid, S. S.; Cowan, J. A., *Biochemistry* **1990**, *29*, 6025–32. (d) Cowan, J. A., *Inorg. Chem.* **1991**, *30*, 2740–47. (e) Black, C. B.; Cowan, J. A., *J. Am. Chem. Soc.* **1994**, *116*, 1174–78.

20. Brandts, J. F.; Lin, L.-N.; Wiseman, T.; Williston, S.; Yang, C. P., *Int. Lab.* **1990**, *20*, 29–35.

21. Moeschler, H. J.; Schaer, J.-J.; Cox, A. J., *Eur. J. Biochem.* **1980**, *111*, 73–78.

22. Bayley, P. M.; Ahlström, P.; Martin, S. R.; Forsén, S., *Biochim. Biophys. Res. Commun.* **1984**, *120*, 185–91.

23. Martin, S. R.; Anderson Teleman, A.; Bayley, P. M.; Drakenberg, T.; Forsén, S. *Eur. J. Biochem.* **1985**, *151*, 543–50.

24. Martin, S. R.; Linse, S.; Bayley, P. M.; Forsén, S., *Eur. J. Biochem.* **1986**, *161*, 595–601.

25. Forsén, S.; Linse, S.; Thulin, E.; Lindegård, B.; Martin, S. R.; Bayley, P. M.; Brodin, P.; Grundström, T., *Eur. J. Biochem.* **1988**, *117*, 47–52.

26. Sandström, J., *Dynamic NMR Spectroscopy*, Academic Press, London, 1982.

27. Tsai, M.-D.; Drakenberg, T.; Thulin, E.; Forsén, S., *Biochemistry* **1987**, *26*, 3635–43.

CHAPTER 3

Metal Substitution as a Probe of the Biological Chemistry of Magnesium Ion

Anton Tevelev and J. A. Cowan

3.1 Introduction

Chapter 2 reviewed the physicochemical methods available for direct study of the biological chemistry of magnesium ion. In principle, the use of transition metal ions or complexes should allow one to use additional techniques to probe metal cofactor chemistry; however, a major concern in their use is the possibility of changing the chemistry of the experimental system in ways that may not be readily apparent. Accordingly, one has to carefully examine the relevance of the results obtained before extrapolating any conclusions to the understanding of enzyme activity with the natural magnesium cofactor. In particular, close attention must be paid to possible changes in the location of the metal binding site, coordination geometry, mode of substrate binding, and the mechanism of action. With careful consideration of these factors, such strategies have provided useful insight on the structural and mechanistic roles of magnesium cofactors in biology.

Given the immensity of the task, no effort has been made to present a comprehensive review of every example of metal substitution as applied to the understanding of magnesium biochemistry. Rather, we shall attempt to give the reader a general overview of the great variety of physical and chemical techniques available, and a comprehensive listing of original research papers that fully describe both the background to the technique and its application. Each approach is illustrated by specific examples. Not all of these concern magnesium biochemistry per se; however, with appropriate modifications they could readily be applied to the study of magnesium-dependent molecules and metabolic pathways.

3.2 Metal Substitution—Spectroscopic Methods

The chemistry of only a few metal ions show any close resemblance to the chemistry of divalent magnesium (Table 3.1), and so these metals have been used most commonly as probes in studies of magnesium biochemistry. Divalent manganese shows the greatest similarity in ligand preference and geometry, and is a functional substitute for Mg^{2+} in many biological systems.[1,2] However, the ligand exchange rate (k_{ex}) is 100 times faster than that of Mg^{2+}, which may influence the mechanistic pathway in certain instances.[3] Also, Mn^{2+} shows a higher affinity for nitrogen ligands, which is reflected in the larger number of available binding sites in comparison to Mg^{2+} and can complicate its use as a probe ion.[2] Other ions commonly used as probes for Mg^{2+} in studies of magnesium-dependent enzymes include Co^{3+} and Cr^{3+},[1,4] which are mainly used as complexes with nucleotide di- and triphosphates. Use of divalent nickel as a probe for magnesium has also been reported.[5] Strong similarities also exist between Mg^{2+} and Ca^{2+}, and between Mg^{2+} and Li^{+}[2,6]; however, the spectroscopic silence of these ions reduces their utility as probes.

Whichever metal is eventually selected must be shown to bind specifically at the magnesium binding site, and to maintain a similar ligand coordination set and geometry as magnesium ion. If mechanistic studies are involved, it should be demonstrated that the reaction pathway is not altered by metal substitution, a matter that is not always straightforward.

Some of the most powerful methods for investigations of protein structure and function are spectroscopic in nature: namely, NMR, EPR, and electronic absorption.[a] Unfortunately, divalent magnesium is not generally suited for study by these methods. For this reason, with the minor exception of a limited number of ^{25}Mg NMR experiments (see Chapter 2), an investigator is often faced with the need to substitute Mg^{2+} with a spectroscopic probe ion. In this section, we detail the probe ions that are appropriate for each spectroscopic method, and offer specific examples of their use. This has been done in a manner that demonstrates the strategies to be employed, and the manner in which specific structural and functional information may be derived.

3.2.1 NMR Spectroscopy

Nuclear magnetic resonance spectroscopy is one of the most informative methods available for investigation of biological molecules, being challenged only by X-ray diffraction techniques in the crystalline state. Certain nuclei possess a nonzero spin

---

[a] Abbreviations: ADP, adenosine-5′-diphospahate; cAMP, cycloadenosinemonophosphate; AMPPCP, β,γ-methylene-ATP; ATP, adenosine-5′-triphosphate; CTP, cytosine-5′-triphosphate; DTPA, diethylenetriaminepentaacetic acid; EDTA, ethylenediaminetetraacetic acid; EPR, electron paramagnetic resonance; ESEEM, electron spin-echo envelope modulation; Glu, glutamate; NMR, nuclear magnetic resonance; Phe, phenylalanine; PRE, proton relaxation enhancement.

Table 3.1 PHYSICOCHEMICAL PROPERTIES OF SOME ALKALINE EARTH AND TRANSITION METALS

Ion	Ionic radius (Å)	Charge–radius ratio	k_{ex} (s⁻¹)	Coordination number	Geometry	Ligand preference	Nuclear spin	Electron spin
Mg²⁺	0.86	2.33	5×10^5	6	oct	O	$\frac{5}{2}$	0
Mn²⁺ (HS d^5)	0.97	2.07	2×10^7	6	oct	O, N	$\frac{5}{2}$	$\frac{1}{2}$
Co²⁺ (HS d^7)	0.72	2.78	2×10^6	4, 6	tet, oct	N, S	$\frac{7}{2}$	$\frac{1}{2}$
Co³⁺ (LS d^6)	0.69	4.35	$<10^{-12}$	6	oct	O, N	$\frac{7}{2}$	0
Cr³⁺	0.76	3.95	5×10^{-7}	6	oct	O, N	$\frac{3}{2}$	$\frac{1}{2}$
Ni²⁺	0.69	2.90	3×10^4	6	oct	O, N, S	0	0
Zn²⁺	0.74	2.70	7×10^7	4	tet	N, S	$\frac{5}{2}$	0
Cd²⁺	0.92	2.17	4×10^8	4	tet	N, S	$\frac{1}{2}$	0
Ca²⁺	1.14	1.75	8×10^8	6–8	flexible	O	$\frac{7}{2}$	0
Ln³⁺	1.0–1.2	2.73	2×10^8	6–9	flexible	O		$\frac{1}{2}$

Adapted in part from Refs. 1 and 2.

(*I*), and therefore an intrinsic magnetic moment. In an external magnetic field (usually directed along the *z* axis), the degeneracy of various magnetic states (defined by the orientation of the nuclear magnetic moment) is removed (Fig. 3.1). The energies of transitions between these states fall in the radiowave region of the electromagnetic spectrum, and NMR spectra reflect transitions between these nuclear energy states. The energy required to promote these transitions is usually expressed in terms of the chemical shift (δ) in ppm (parts per million) as defined by

$$\delta = \frac{v - v_s}{v_0} \, 10^6 \qquad\qquad (3.1)$$

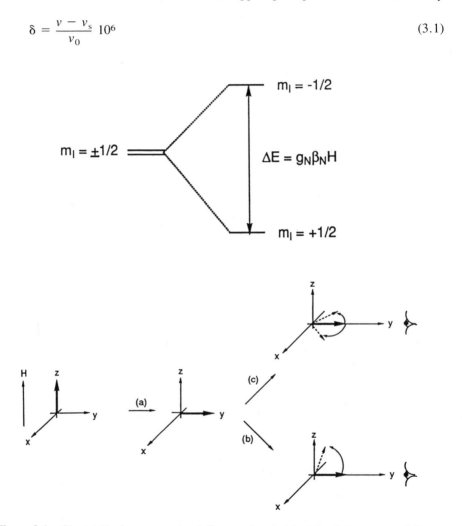

Figure 3.1 (Upper) Nuclear energy-level diagram showing the loss of degeneracy of the m_I nuclear spin states in a magnetic field *H*. The resonance frequencies described in the text refer to the energy gap $\Delta E = hv$. (Lower) In a bulk sample the individual spin vectors can be considered as an overall magnetization. By application of a magnetic field the magnetization vector can be transformed into the *xy* plane for detection (a). The magnetization decays as the spin vectors realign along the *z* axis at a rate defined by: (b) the longitudinal relaxation time (T_1), and (c) by loss of coherence in the *x–y* plane defined by the relaxation time (T_2).

where v is the radio frequency of the transition promoted in the sample, v_s is the frequency taken up by a standard, and v_o is the fixed frequency of the probe.[7] The value of the chemical shift depends on the local magnetic field surrounding a nucleus and reflects the local structure of a molecule. Furthermore, a nucleus in an excited state will spontaneously return to the ground state with a rate characterized by two independent times (Fig. 3.1): the longitudinal relaxation time (T_1) and the transverse relaxation time (T_2). T_1 describes the relaxation rate of the total magnetization along the z axis, while T_2 reflects the loss of coherence of individual spins in the $x–y$ plane. Relaxation mechanisms may involve dipole–dipole or contact interactions with other nuclei, anisotropy in the local magnetic field, or interactions with unpaired electrons.[8] Both T_1 and T_2 can be measured relatively easily, and many structural or dynamic factors of interest to the biochemist may influence their values, including the size of a molecule, chemical exchange rates, and the distance to a paramagnetic nucleus.[8,9]

NMR experiments are best carried out on nuclei with spin $I = \frac{1}{2}$. Quadrupolar nuclei (nuclei with spin greater than $\frac{1}{2}$) often give very broad signals, and are generally unsuitable for NMR experiments. ^{25}Mg has a $\frac{5}{2}$ spin, and as a result ^{25}Mg NMR is technically demanding (Chapter 2). For this reason, Mg^{2+} is often replaced with paramagnetic metals (Co^{2+}, Cr^{3+}, and Mn^{2+}) that perturb the chemical shifts and relaxation times of neighboring nuclei in a protein and/or substrate that are spectroscopically more amenable (^1H, ^{13}C, ^{15}N, ^{31}P).[5]

3.2.1.1 Determination of Distances between Enzyme and Substrates Based on Changes in the Longitudinal Relaxation Rates of NMR-Active Nuclei Due to a Neighboring Paramagnetic Probe

The following equations illustrate how the effect of a paramagnetic probe depends on the distance between the nucleus under observation and the paramagnetic metal ion:

$$\frac{1}{T_{1M}} = \frac{2}{15} \frac{S(S+1)\gamma_I^2 g^2 \beta^2}{r^6} \left(\frac{3\tau_c}{1 + \omega_I^2 \tau_c^2} + \frac{7\tau_c}{1 + \omega_S^2 \tau_c^2} \right) \quad (3.2)$$

$$+ \frac{2}{3} \frac{S(S+1)A^2}{\hbar^2} \left(\frac{\tau_e}{1 + \omega_S^2 \tau_e^2} \right)$$

$$\frac{1}{T_{2M}} = \frac{1}{15} \frac{S(S+1)\gamma_I^2 g^2 \beta^2}{r^6} \left(4\tau_c + \frac{3\tau_c}{1 + \omega_I^2 \tau_c^2} + \frac{13\tau_c}{1 + \omega_S^2 \tau_c^2} \right)$$

$$+ \frac{1}{3} \frac{S(S+1)A^2}{\hbar^2} \left(\tau_e + \frac{\tau_e}{1 + \omega_S^2 \tau_e^2} \right)$$

Here $1/T_{1M}$ and $1/T_{2M}$ are the paramagnetic contributions to the longitudinal and transverse relaxation rates of a molecule bound near the paramagnetic nucleus, S is the electron spin, γ_I is the nuclear magnetogyric ratio, g is the electronic g factor, β is the Bohr magneton, r is the metal–nucleus distance, ω_I and ω_S are the nuclear and electron Larmor precession frequencies, τ_c and τ_e are the correlation times for the dipolar and contact interactions, and iA is the contact hyperfine coupling constant.[10] The dipolar term of the Solomon–Bloembergen equation (3.3) can be used to calculate the distances between a paramagnetic center and a nucleus (provided the relaxation rates are determined), where the parameter B is a collection of constants and is defined by equation (3.4)

$$r(A) = B[qT_{1M}f(\tau_c)]^{1/6} \tag{3.3}$$

$$B = [\tfrac{2}{15}S(S+1)\gamma_I^2 g^2 \beta^2]^{1/6} \tag{3.4}$$

(for example, for the Cr^{3+}–proton interactions, B is equal to 705),[11] T_{1M} is the relaxation time of a nucleus, q is the coordination number, and the correlation function $f(\tau_c)$ is defined by Eq. (3.5). These equations allow one to estimate distance constraints between a molecule and a bound-metal ion. Note that the correlation time (τ_c) is best defined by Eq. (3.6),[11] which offers more insight on the physical parameters that characterize this factor, where τ_r is the time constant for rotation of the paramagnetic complex, τ_s is the electron spin relaxation time, and τ_r is the residence time in the bound state.

$$f(\tau_c) = \frac{3\tau_c}{1 + \omega_I^2 \tau_c^2} + \frac{7\tau_c}{1 + \omega_s^2 \tau_c^2} \tag{3.5}$$

$$(\tau_c)^{-1} = (\tau_r)^{-1} + (\tau_s)^{-1} + (\tau_m)^{-1} \tag{3.6}$$

This approach has been used to determine distances from the Mn^{2+} site in *E. coli* RNA polymerase to a nucleotide in the initiation site, and between nucleotides in the initiation and elongation sites (Fig. 3.2).[12] It was shown that at the concentrations of substrates employed, the elongation site was occupied only by MgCTP and CrATP occupied the initiation site, forming a quaternary enzyme–CrATP–MgCTP complex. The proton NMR spectrum of MgCTP in the enzyme complex was measured in the absence and in the presence of CrATP, and the longitudinal relaxation rates of three easily resolvable protons (H1′, H5, and H6; Fig. 3.2) of CTP were determined. Figure 3.3 further illustrates the methodology for the binding of adenylyl(3′-5′)uridine to RNA polymerase.[12] The paramagnetic effects of enzyme-bound CrATP on the relaxation rates of the protons of CTP $[(1/fT_{1P})_{enz}]$ were calculated (Table 3.2a) by use of

$$\frac{1}{T_{1P}} = \frac{[CrATP]_{bound}}{[CTP]_{total}} \left(\frac{1}{fT_{1P}}\right)_{enz} + \frac{[CrATP]_{free}}{[CTP]_{total}} \left(\frac{1}{fT_{1P}}\right)_{free} \tag{3.7}$$

where $1/T_{1P}$ is the difference in $1/T_1$ values determined in the presence and the absence of CrATP, f is a normalization factor equal to $[CrATP]_{bound}/[CTP]$, and $(1/fT_{1P})_{free}$ is the normalized paramagnetic effect of free CrATP, $(1/fT_{1P})_{free}$ is the normalized paramagnetic effect of free CrATP, and $(1/fT_{1P})_{enz}$ is the unknown to be determined, being the paramagnetic effect of enzyme-bound CrATP on the relaxation rates of the protons of CTP. Equation (3.3) can be used to calculate distances

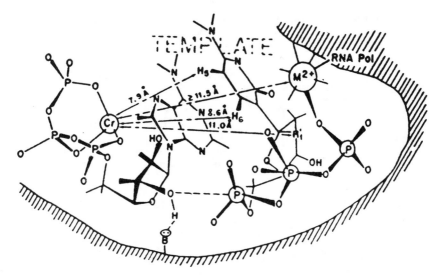

Figure 3.2 *E. coli* RNA polymerase. CrATP is located at the initiation site, and CTP is at the elongation site. Adapted from Ref. 13.

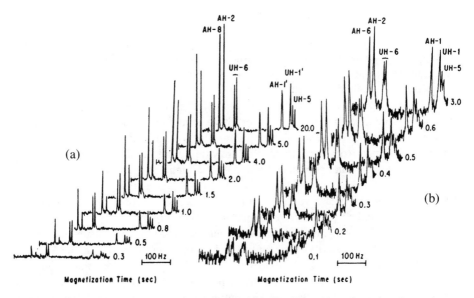

Figure 3.3 ^{1}H NMR spectrum of free adenylyl(3'-5')uridine (a), or bound to the paramagnetic Mn(II)ATP *E. coli* RNA polymerase complex (b). The AH8, AH2, UH6, AH1', UH1', and UH5 protons were closely monitored. Adapted from Ref. 12.

Table 3.2 EVALUATION OF INTERSUBSTRATE DISTANCES ON *E. coli* RNA POLYMERASE BY MAGNETIC RESONANCE MEASUREMENTS

(a) Paramagnetic effects of the RNA polymerase–CrATP complex on the longitudinal relaxation rates of the protons of CTP at 100 MHz[a]

Proton	[CrATP] (μM)	$1/T_1$ (s^{-1})	$1/T_{1P}$ (s^{-1})	$1/(fT_{1P})$ (Enz) (s^{-1})
H-6	0	1.79	—	—
H-6	100	3.33	1.54	225
H-5	0	1.35	—	—
H-5	100	3.13	1.78	357
H-1'	0	1.54	—	—
H-1'	100	2.22	0.68	114

[a] The enzyme concentration was 77.8 μM.

Adapted from Ref. 13.

(b) Calculation of distances from CrATP to the protons of CTP on RNA polymerase

Proton	$1/fT_{1P}$ (s^{-1})	$f(\tau_c)$ (s $\times$ 10^9)	r (Å)
H-6	337	1.1	8.6
H-5	588	1.1	7.9
H-1'	80	1.1	11.0

Adapted from Ref. 13.

from a paramagnetic metal ion to a ligand only if the $1/fT_{1P}$ values are not limited by ligand exchange rates to the enzyme, otherwise only upper limits for the intersite distances can be calculated, where $T_{1M} = (fT_{1P})_{enz}$. In the case of RNA polymerase it was demonstrated that relaxation of the protons on CTP was not limited by chemical exchange. To calculate the correlation function $f(\tau_c)$ by use of Eq. (3.5), the correlation time for the dipolar interactions (τ_c) must be known. This parameter can be determined from the influence of enzyme-bound CrATP on the longitudinal relaxation rates of water protons obtained over a range of applied frequencies. The best experimental value, $\tau_c = (4.0 \pm 2.0) \times 10^{-10}$ s, gave $f(\tau_c) = (1.13 \pm 0.54) \times 10^{-9}$ s at 100 MHz. These values, and the averaged values of $1/fT_{1P}$, were used in Eq. (3.3) to calculate the distances noted in Table 3.2b.[13]

In a similar fashion, intersite distances between the two bound substrate molecules of pyruvate kinase (pyruvate and MgATP) were determined using CrATP as a MgATP analog (Fig. 3.4).[11] By use of ^{13}C-enriched pyruvate, ^{13}C relaxation rates were determined in the absence and in the presence of CrATP. These studies were carried out at pH 6.6 since the CrATP complex was found to be unstable at higher pH, while the lower pH enhanced deuteration of the pyruvate methyl group (Table 3.3a). The normalized paramagnetic contribution to the relaxation of ^{13}C nuclei by

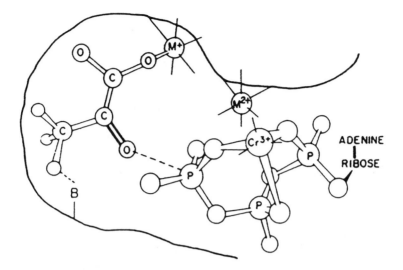

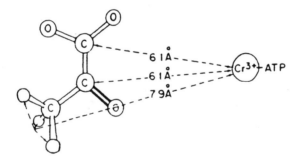

Figure 3.4   Active site of pyruvate kinase. Adapted from Ref. 11.

the enzyme-bound CrATP was calculated by use of Eq. (3.7), here considering pyruvate rather than CTP. The correlation times (τ_c) for CrATP nuclear dipolar interactions required for Eqs. (3.3) and (3.5) were determined from the frequency dependence of the water proton relaxation rates in a variety of CrATP complexes (Table 3.3b) by use of Eqs. (3.3) and (3.5), and assuming an electron–proton distance $r = 2.8$ Å. In all cases approximately 50% of the CrATP present was bound to the enzyme, and data were corrected for occupancy to obtain the water proton relaxation rates due to enzyme-bound CrATP. In this way correlation times of $(1.2–2.3) \times 10^{-10}$ s were determined. The electron spin relaxation time of CrATP was determined by EPR spectroscopy as 10^{-10} s.

   With these data in hand, the distances between the paramagnetic Cr^{3+} ion and

Table 3.3 EVALUATION OF INTERSUBSTRATE DISTANCES ON PYRUVATE KINASE BY MAGNETIC RESONANCE MEASUREMENTS

(a) Effect of CrATP on ^{13}C longitudinal relaxation rates of carbonyl and carboxyl carbon atoms of pyruvate at 25.1 MHz

[Enzyme sites] (mM)	[CrATP] (mM)	[CrATP]$_b$ (mM)	Carbonyl ^{13}C			Carboxylate ^{13}C		
			$(T_1)^{-1} \times$ 10^2 (s^{-1})	$(T_{1P})^{-1} \times$ 10^2 (s^{-1})	$(fT_{1P})_b^{-1}$ (s^{-1})	$(T_1)^{-1} \times$ 10^2 (s^{-1})	$(T_{1P})^{-1} \times$ 10^2 (s^{-1})	$(fT_{1P})_b^{-1}$ (s^{-1})
0.400	—	—	3.3	—	—	2.2	—	—
0.395	0.11	0.068	11.8	8.4	74	10.0	7.8	68
0.380	0.38	0.187	25.0	21.7	68	28.6	26.3	83
0.076	0.38	0.047	7.7	5.4	59	7.1	5.6	62

Adapted from Ref. 11.

(b) Frequency dependence of normalized longitudinal relaxation rate $(1/fT_{1P})$ of water protons and correlation times (τ_c)

Complex	$(1/fT_{1P}) \times 10^{-5}$ (s^{-1})				$\tau_c \times 10^{10}$ (s)
	3 MHz	8 MHz	15 MHz	24.3 MHz	
CrATP	4.7	3.2	3.0	3.0	1.4
CrATP (+ pyruvate)	3.3	2.2	2.0	1.9	1.2
E·Mg·CrATP	6.4	5.3	5.1	5.1	2.3
E·Mg·CrATP (+ pyruvate)	5.3	3.8	3.6	3.5	1.5

Adapted from Ref. 11.

(c) Distances between Cr^{3+} of enzyme-bound CrATP and carbonyl and carboxyl carbon atoms and methyl protons of pyruvate

Nucleus	$(1/fT_{1P})_b$	$\tau_c \times 10^{10}$ (s)	Distance (Å)
Methyl protons	218	1.5	7.9
Carbonyl carbon	67	1.5	6.1
Carboxyl carbon	71	1.5	6.1

Adapted from Ref. 11.

diamagnetic nuclei (^{13}C) on the substrate were calculated from the paramagnetic contribution to the longitudinal relaxation rates. Again, it was demonstrated that these rates were not limited by chemical exchange and were not dominated by outer sphere relaxation (Table 3.3c). Averaged values of $1/(T_{1P})_b$ were used with Eq. (3.3) to calculate intersite distances [note that for Cr(III)–^{13}C interactions, $B = 445$), and were confirmed by analyzing the influence of a Mn^{2+} probe ion on ^{31}P relaxation rates.[12]

3.2.1.2 Determination of a Metal Binding Constant by Proton Relaxation Rate Enhancement

The presence of a bound metal complex in the vicinity of a second solvated paramagnetic metal ion may influence the longitudinal relaxation rates of the protons in the water molecules bound to the second paramagnetic cation.[14] By such methods the dissociation constant of Mn^{2+} for the cAMP-dependent bovine heart muscle protein kinase complex with bound substrate has been determined by PRE methods.[15] Typically, Mn^{2+} binds at an inhibitory site on the enzyme and MgATP is the substrate. In the PRE experiments, cobalt or chromium complexes of the ATP analog β,γ-methylene-ATP (AMPPCP) were used. Preliminary studies demonstrated that the substrate analogs bind to the same site and in the same manner as MgATP. Subsequently, a mixture of enzyme and (NH$_3$)$_4$CoAMPPCP was titrated with Mn^{2+}(aq), and the relaxation rates of water protons were measured. From these data, the ratios of free and enzyme-bound Mn^{2+} were determined,[15,16] and the results presented in the form of a Scatchard plot (Fig. 3.5) to yield 1.25 ± 0.20 tight

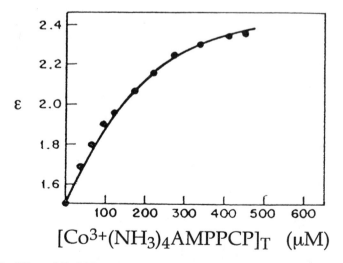

Figure 3.5 Effect of Co(NH$_3$)$_4$AMPPCP binding to protein kinase (50.4 μM) in the presence of Mn^{2+} (101 μM), on the enhancement of the proton relaxation rate of bulk water. Adapted from Ref. 15.

Mn^{2+} binding sites with $K_D \sim 172 \pm 20$ μM. Similar studies using CrAMPPCP yielded an estimate of 1.1 ± 0.1 tight Mn^{2+} sites with $K_D \sim 55 \pm 10$ μM. It proved possible to monitor direct binding of the paramagnetic CrAMPPCP complex by PRE methods, even in the absence of paramagnetic Mn^{2+}. With CrAMPPCP, a dissociation constant of 25 ± 10 μM and an enhancement factor of 1.4 ± 0.2 were obtained. In the presence of Mg^{2+} the dissociation constant was lowered to 8 ± 3 μM.

3.2.1.3 Determination of Intermetal Distances by the Proton Relaxation Method

The distance between two paramagnetic metals can be determined in a similar fashion. The $Cr^{3+}(NH_3)_4ATP$ complex can be used as a paramagnetic analog of MgATP, and Mn^{2+} as a probe for an enzyme-bound Mg^{2+} site. In this way, the distance between a divalent metal site in the catalytic center of the rabbit muscle pyruvate kinase and MgATP was determined.[15,16] Control data were obtained with the diamagnetic Co(III)ATP complex and Mg^{2+}, respectively. The interested reader should refer to Refs. 13 through 16 for further details.

3.2.2 EPR Spectroscopy

The theory of electron paramagnetic resonance spectroscopy (EPR) is very similar to that of NMR. In the EPR experiment, one monitors transitions between the energy levels available to an unpaired electron. For technical reasons, an EPR spectrum is usually presented as the first derivative of an absorption curve versus the strength of the magnetic field (in gauss). A dimensionless g value is reported, defined by

$$g = \frac{h\nu}{\beta H} \tag{3.8}$$

where ν is the fixed frequency of the probe, H is the magnetic field strength, and β is the electron Bohr magneton ($\beta = 9.2741 \times 10^{-24}$ J T^{-1}). As the examples to follow demonstrate, the EPR signature of a paramagnetic metal ion is not as sensitive to the chemical environment as the chemical shift in NMR, but can be usefully employed. Since only paramagnetic metal ions and organic free radicals are EPR active, $Mn^{2+}(aq)$ or $Cr^{3+}ATP$ analogs are often used to probe Mg^{2+} binding sites.

3.2.2.1 Determination of Metal-Enzyme Dissociation Constants.

EPR can be used as a simple analytical tool to determine the relative concentrations of free and enzyme-bound paramagnetic cations during a titration study.[17] Typically, Mn^{2+} is employed since the enzyme-bound ion is EPR silent. The binding of Mn^{2+}

to the complex of bovine heart muscle protein kinase and $Co(NH_3)_4$-β,γ-methylene-ATP has been studied in this way,[15] and a Scatchard plot revealed 1.25 ± 0.20 high-affinity Mn^{2+} sites with $K_D \sim 170 \pm 20$ μM (Fig. 3.6).

3.2.2.2 Determination of Coordination Geometry Based on EPR Spectroscopy

EPR spectra of high-spin Co^{2+} complexes generally give three g values that reflect the ligand symmetry. However, in practice, it is often difficult to distinguish between tetrahedral, pentacoordinate, and octahedral Co^{2+} complexes. Nevertheless, hyperfine coupling of the unpaired electrons with the ^{59}Co nucleus is usually larger in five- and six-coordinate complexes than in tetrahedral complexes, resulting in well-defined splittings or broader resonances at low field.[18,19]

The proofreading $3',5'$-exonuclease activity of *E. coli* DNA polymerase I requires divalent cation cofactors such as Mg^{2+}, Mn^{2+}, or Zn^{2+}. The geometry of the metal binding sites have been studied following substitution with a paramagnetic probe ion. The enzyme is active with bound Co^{2+}, and the EPR spectra of the Co^{2+}–$3',5'$–exonuclease complex in the absence and presence of TMP and Zn^{2+} show a broad low-field resonance, suggesting a large hyperfine coupling constant. From this evidence it has been concluded that the cobalt ion is five- or six-coordinate.[20] Addition of excess Co^{2+} led to an increase in the intensity of the EPR

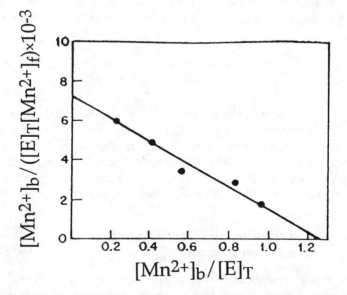

Figure 3.6 Scatchard plot of Mn^{2+} binding to the enzyme–$Co(NH_3)_4$AMPPCP complex. The concentrations of the enzyme and $Co(NH_3)_4$AMPPCP are 125.7 and 324 μM, respectively. Adapted from Ref. 15.

spectrum, but obscured the hyperfine splitting due to nonspecific binding of Co^{2+} (Fig. 3.7).

3.2.2.3 Determination of Coordination Environment Uing ESEEM

If an appropriate sequence of microwave pulses is applied to a paramagnetic sample, a spin echo is observed as a result of refocusing by the unpaired electrons. When the pulse sequence is incremented in time, a time-dependent sequence of spin echoes is obtained. The amplitudes of the resulting spin echoes generate an envelope. If a nuclear spin is coupled to the electron spin, the envelope of echoes is modulated in a periodic manner by the nuclear spin, and so gives rise to electron spin-echo envelope modulation (ESEEM) spectroscopy.[21-24] This modulation varies in frequency, amplitude, duration, and depth, depending on the magnetogyric ratios of the ligand nuclei, their number, metal–ligand distance, magnitude of the coupling constant to the electron spin, and the magnitude of the nuclear spin. For example, the modulation depth for the deuteron is $\frac{8}{3}$ that of a proton, while the Larmor precession frequency is five-fold lower.[25]

Determination of the number of water molecules coordinated to a metal cofactor is often a critical factor in evaluating the ligand composition and geometry of the metal site. If the natural metal cofactor can be substituted by Mn^{2+}, then ESEEM can be used. As an illustration of the method, the ESEEM spectra of two model

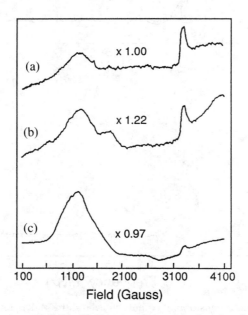

Figure 3.7 EPR spectra of $Co^{2+}-3',5'$-exonuclease complexes at 8 K: (a) 151 μM enzyme, 151 μM Co^{2+}; (b) as in (a) with 302 μM TMP; (c) 151 μM enzyme, 755 μM Co^{2+}, and 151 μM Zn^{2+}. Adapted from Ref. 20.

compounds that differ by only a single bound water molecule [Mn^{2+}–EDTA and Mn^{2+}–DTPA (diethylenetriaminepentaacetic acid)] have been examined.[26] Differences in the ratio of ESEEM spectra of these compounds obtained in H_2O and D_2O arise solely from exchangable deuterium. In this case, the ratio of the spectra of Mn^{2+}–EDTA and Mn^{2+}–DTPA reflects a single water molecule in the inner sphere of Mn^{2+} (Fig. 3.8). The method has also been applied to studies of metalloproteins. In the Ca^{2+}-dependent staphylococcal nuclease, the number of water molecules coordinated to Mn^{2+} (replacing Ca^{2+}) in the binary complex of the enzyme with Mn^{2+}, and in the ternary enzyme–Mn^{2+}–3′,5′-pdTp complex were estimated.[27] Comparison of spectra obtained in H_2O and D_2O discriminated contributions from ^{14}N, ^{31}P, and nonexchangeable protons in the vicinity of the paramagnetic center

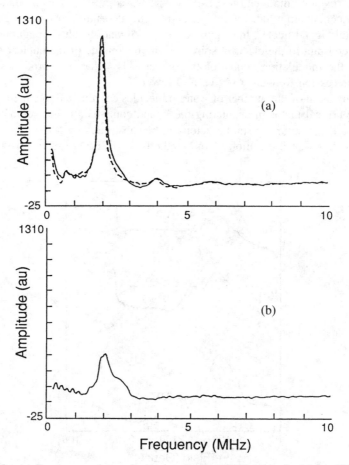

Figure 3.8 (a) Fourier transform of stimulated-echo ESEEM data for Mn^{2+}–EDTA (solid line) and Mn^{2+}–DPTA (dashed line). (b) Fourier transform of the difference between the stimulated-echo ESEEM data for Mn^{2+}–EDTA and Mn^{2+}–DPTA. All spectra were measured in ^2H_2O. Adapted from Ref. 27.

(Fig. 3.9). A comparison of ratioed data for protein complexes in D_2O and H_2O showed that the ternary complex of the enzyme with Mn^{2+} and 3′,5′-pdTp had a more intense deuterium modulation in the ratioed data than did the binary complex. The ratio of deuterium modulation for ternary and binary complexes gave rise to a line at the deuterium Larmor frequency, the intensity and the width of which were consistent with results expected for a single water molecule coordinated to Mn^{2+}. Thus the ternary complex appeared to have one additional water of hydration relative to the binary complex, provided that contributions from ambient water were identical for both complexes and that water protons were the major exchangeable species close to Mn^{2+}.

3.2.2.4 Intermetal Distances

A paramagnetic metal ion may influence the transverse electron-spin relaxation time of electrons on an adjacent paramagnetic center. *E. coli* glutamine synthetase binds a simple divalent metal ion, but also binds a second ion as a metal–nucleotide complex. The enzyme·Mn^{2+}·sulfoximine complex has been used to determine the intermetal distance to Cr(III)ATP (Fig. 3.10).[27a] Addition of Cr(III)ATP resulted in a drastic reduction of the amplitude of the enzyme-bound Mn^{2+} EPR signal, without affecting its linewidth. At saturating levels of Cr(III)ATP, the amplitude of

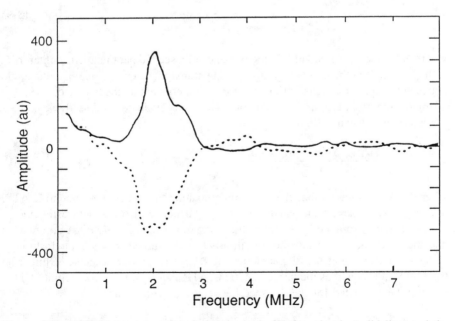

Figure 3.9 ESEEM spectrum attributed to a single ^2H_2O molecule (solid line) and the difference ESEEM spectrum between the binary enzyme–Mn^{2+} complex, and the ternary enzyme–Mn^{2+}–pdTp complex (dashed line). Adapted from Ref. 27.

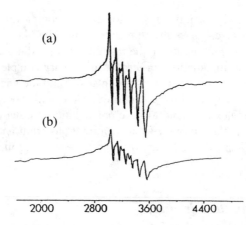

Figure 3.10 Effects of Cr(III)ATP· binding on the EPR spectrum of the ternary glutamine synthetase–Mn^{2+}–L-methionine-($\pm$)-sulfoximine complex obtained at 9 GHz. Spectrum (a): 0.3 mM glutamine synthetase, 0.25 mM Mn^{2+}, 6 mM sulfoximine. Spectrum (b): spectrum (a) plus 1.3 mM Cr(III)ATP. Adapted from Ref. 27a.

individual transitions decreased 85%. The interaction coefficient (C) was calculated from:

$$C = \frac{\delta H - \delta H_0}{1 - 3\cos^2\theta'_R} \tag{3.9}$$

where δH is the EPR linewidth in the presence of a second paramagnetic center, δH_0 is the linewidth in the absence of the second paramagnetic center, and θ'_R is the angle between the applied magnetic field and the vector connecting the two electron spins. Using the theory of Leigh,[27b] a value of $C = 40$ was evaluated and used to calculate the intermetal distance (r) according to

$$C = \frac{(g)_{Mn}[g^2S(S+1)]_{Cr}\,\beta^3\tau_{Cr}}{\hbar r^6} \tag{3.10}$$

where τ_{Cr} is the longitudinal electron-spin relaxation time of enzyme-bound CrATP, and γ, β, and S have their usual meanings. A distance of 6.8 Å was estimated.

The distance from the Mn^{2+} ion at the elongation site of *E. coli* RNA polymerase and the ATP-bound Cr^{3+} at the initiation site was studied in a similar fashion.[13] However, inasmuch as the diamagnetic Co(NH$_3$)$_4$ATP produced the same effect on the Mn^{2+} EPR signal as the paramagnetic CrATP analog, only a lower limit of 11.5 Å could be estimated for the distance between the metal centers.

3.2.3 Electronic Absorption Spectroscopy

The energy difference between two electronic states typically lies in the range of ultraviolet or visible light. Absorption of electromagnetic radiation in this energy range by an ion or molecule can stimulate an electronic transition from a low- to a high-energy electronic state. The difference in intensity between the incident (I_o) and transmitted (I_t) radiation is measured, and the absorbance (A) defined in the following equation is plotted as a function of the irradiation wavelength.

$$A = \log_{10}(I_o/I_t) \tag{3.11}$$

For transition metal ions, electronic transitions typically differ in the arrangement of electrons in d orbitals (d–d transitions). The difference in energy depends on the ligand-field splitting of the d orbitals, which in turn depends on the chemical nature of the ligands and their geometry around the metal ion.[28] Magnesium ion has no d electrons and consequently shows no electronic transitions in the visible or near-UV regions. However, Co^{2+} shows very characteristic UV–visible spectra that can be useful probes of the geometry of a metal-binding site.

3.2.3.1 Determination of the Coordination Environment from Optical Spectra

Metal ions that show structure-dependent electronic spectra can be of value in determining the coordination geometry of a metal cofactor when bound at the active site. Divalent cobalt is a useful probe ion for this purpose. Ligand-field theory predicts that optical transitions of four-coordinate or tetrahedral Co^{2+} complexes will give rise to intense absorptions ($\epsilon > 300$ M^{-1} cm^{-1}) around 625 ± 50 nm, while transitions of six-coordinate or octahedral Co^{2+} complexes show absorption bands with smaller extinction coefficients ($\epsilon < 30$ M^{-1} cm^{-1}) at shorter wavelengths (525 ± 50 nm). Pentacoordinate Co^{2+} complexes show moderate absorption ($50 < \epsilon < 250$ M^{-1} cm^{-1}) with several maxima between 525 and 625 nm.[29,30]

The proofreading 3′,5′-exonuclease activity of the Klenow fragment of DNA polymerase I requires divalent cations such as Mg^{2+}, Mn^{2+}, or Zn^{2+} for functional activity.[31] It has been shown that Co^{2+} stimulates enzyme activity,[20] and binds to the Klenow fragment in a cooperative fashion with a Hill coefficient of 2.3 at Co^{2+} concentrations of ≤ 8 μM, and with a Hill coefficient of 1.2 at higher Co^{2+} concentrations, with an apparent dissociation constant of 16.6 μM. The same cooperative behavior was observed for Mg^{2+} and Mn^{2+} ion. It was concluded that at least three divalent metal binding sites were required. Divalent manganese was found to bind most tightly, Mg^{2+} least tightly, and Co^{2+} bound with intermediate affinity. The concentrations of metal ions that gave half-maximum velocity were as follows: $K_{0.5} = 4.2$ μM for Mn^{2+}; $K_{0.5} = 16.6$ μM for Co^{2+}; and $K_{0.5} = 343$ μM for Mg^{2+}. Binding of Co^{2+} caused a marked change in the visible absorption spectrum, with three maxima at 633 nm ($\epsilon = 190$ M^{-1} cm^{-1}), 570 nm ($\epsilon = 194$ M^{-1} cm^{-1}), and 524 nm ($\epsilon = 150$ M^{-1} cm^{-1}) [Fig. 3.11a]. A spectrophotometric titration with Co^{2+} (monitoring the absorbance at 633 nm) in the presence of 1 equivalent of TMP revealed a sigmoidal curve. A similar curve was obtained in the absence of TMP. The best fit to the data was obtained with the assumption that three Co^{2+} ions could bind to the enzyme with complete cooperativity (i.e., with a Hill coefficient of 3) and that the enzyme–$(Co^{2+})_3$ complex was the only species spectroscopically detectable at 633 nm (Fig. 3.11(b)). Addition of 1 equivalent of Zn^{2+}

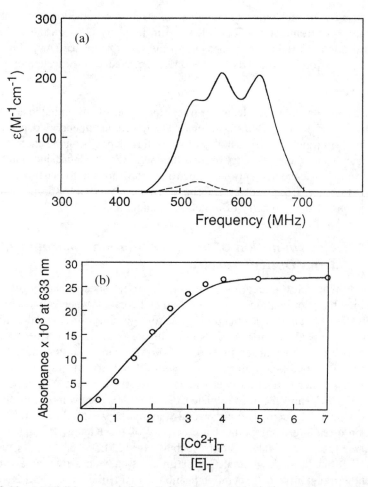

Figure 3.11 (a) Visible spectra of the Co^{2+} complex with 3′,5′-exonuclease obtained with (dashed line) and without (solid line) Zn^{2+} present. (b) Spectrophotometric titration of 3′,5′-exonuclease with Co^{2+} in the presence of TMP, monitoring the absorbance at 633 nm. Adapted from Ref. 20.

in the presence of a sevenfold excess of Co^{2+} resulted in a marked decrease in the intensity of the absorbance ($\epsilon \sim 30$ M^{-1} cm^{-1}) and a shift in the absorption maximum to 524 nm. Since it is known that Zn^{2+} has a high affinity for site A, it became clear that the original absorption spectrum was dominated by Co^{2+} bound at this location. Accordingly, it has been suggested that site A is pentacoordinate, and that each of the two remaining sites are hexcoordinate (see Fig. 6.5).

3.2.3.2 Evaluation of Substrate or Ligand Binding

The optical spectrum of Co^{2+} depends both on the chemical nature of the ligands and on the geometry of the complex. Cobalt(II) has been substituted for zinc ion in carboxypeptidase A and thermolysin.[32] With these derivatives it proved possible to demonstrate that an inhibitor containing a sulfhydryl group (HS-Ac-D-Phe) interacts directly with the metal cofactor in the active site. The spectral shifts and the approximate twofold increase in absorbance after addition of the sulfhydryl reagent pointed to a change in either the ligands or coordination geometry (or both) of the cobalt ion in caboxypeptidase A (Fig. 3.12). The new transition at λ_{max} = 338 nm (ϵ = 900 M^{-1} cm^{-1}) was assigned to a sulfur → cobalt(II) charge-transfer band.[33] Titration experiments with the inhibitor showed that the maximum increase in the charge-transfer band was reached at an equimolar ratio of inhibitor to substrate. Addition of an excess of known competitive inhibitors eliminated this band, indicating that HS-Ac-D-Phe was bound directly to the active site.

3.3 Metal Substitution—Chemical Methods

Most complexes of magnesium ion are kinetically labile. To distinguish between mechanistic intermediates it is often useful to substitute Mg^{2+} for a probe metal ion with a slower exchange rate. In this way it may be possible to distinguish between different diastereomeric forms of metal–substrate or metal–product complexes. An additional advantage that such a strategy offers is the ability to investigate the protein environment inside a metal binding pocket, and to examine the role of metal-bound water molecules. Metal ions that form inert complexes should retain the complete ligand set over the time scale of the experiment.

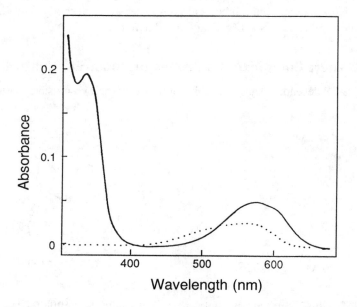

Figure 3.12 Electronic absorption spectra of Co^{2+}–carboxypeptidase A in the absence (dotted line) and in the presence (solid line) of the inhibitor HSAc–D-Phe. Adapted from Ref. 33.

3.3.1 Ion Complexes as Probes of Catalytic Mechanisms

A variety of substitutionally inert cobalt(III) and chromium(III) complexes are available for use as mechanistic probes of labile divalent magnesium, which is always present in solution as an aquated ion. Topoisomerase I requires magnesium as a metal cofactor; however, it has been demonstrated that a variety of inert cobalt(III) complexes (below) also activate the enzyme.[34] The extent of activation depends, in part, on the total charge of the cobalt complex. Systematic variation of the metal-bound ligands affords a method of analyzing the contributions of electrostatics and hydrogen bonding to interactions with enzyme and substrate.[35]

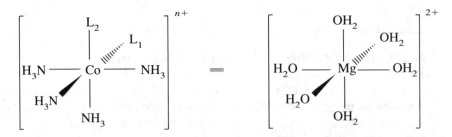

Mg^{2+}-dependent enzymes can function by way of an inner- or an outer-sphere mechanism. To distinguish between them, the use of inert Co^{3+} complexes has been demonstrated.[36] For example, cobalt(III)hexaammine ($L_1 = L_2 = NH_3$) activates the Mg^{2+}-dependent enzyme ribonuclease H from *E. coli*. This complex is similar in size to Mg$(H_2O)_6^{2+}$, but is substitutionally inert, suggesting that the essential metal cofactor does not bind directly to phosphate, and does not provide a nucleophilic water molecule from its primary coordination shell.

3.3.2 Inert Complexes as Probes of Substrate Conformation

Normally ATP-dependent enzymes utilize only one of the two possible diastereomers of MgATP:

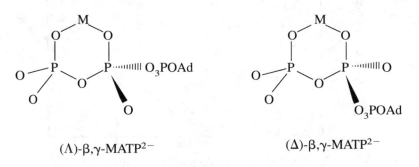

(Λ)-β,γ-MATP^{2-} (Δ)-β,γ-MATP^{2-}

However, interconversion of these two diastereomers in solution is extremely rapid as a result of the lability of Mg^{2+}. The racemic Co(NH$_3$)$_4$ATP substrate mixture displayed a $K_M = 0.65$ mM and $V_{max} = 0.13$ s^{-1}, which was only $\sim$ 0.06% of the V_{max} for the natural MgATP substrate. Both isomers exhibit competitive inhibition versus MgATP. To determine which diastereomer is the actual substrate for the enzyme yeast hexokinase, an inert analog complex [Co(NH$_3$)$_4$ATP] of defined stereochemistry and configuration has been used.[37] Initially a mixture of the two diastereomers was prepared and used as a substrate for the hexokinase.[38] The unused substrate diastereomer was then separated from the product [Co(NH$_3$)$_4$(glucose-6-phosphate)ADP] by use of cycloheptaamylose gel chromatography, and subsequently degraded to Co(NH$_3$)$_4$H$_2$P$_3$O$_{10}$ by successive treatments with periodic acid and aniline. The absolute configuration of the resulting Co(NH$_3$)$_4$H$_2$P$_3$O$_{10}\cdot$H$_2$O complex was determined by X-ray diffraction crystallography,[39] and so it was shown that the actual substrate for the yeast hexokinase had the Λ configuration. Since the enzyme produced cis-Cr(H$_2$O)$_4$ADP(glucose-6-phosphate), the product from the normal reaction with MgATP is cis-Mg(H$_2$O)$_4$ADP(glucose-6-phosphate).[40]

3.4 Substrate Modification

By making careful changes to functional groups on a substrate molecule that putatively make direct contact with the metal cofactor, it may be possible to confirm that the metal cofactor really coordinates to that site.

3.4.1 Phosphorothioates as Probes of Magnesium Binding

Ribozymes catalyze the hydrolysis or transesterification of phosphodiester bonds in ribonucleic acids by attack of water or guanosine, respectively. The reaction usually requires divalent cofactors such as Mg^{2+}.[41,42] Phosphorothioate derivatives of RNA have been used to determine whether the metal cofactor interacts directly with the phosphate oxygen.[43] With Mg^{2+} as a metal cofactor, the phosphorothioate analog is often a poor substrate for the ribozyme. Use of Mn^{2+} or Zn^{2+} rather than Mg^{2+} restored the activity of the ribozyme [Table 3.4, Fig. 3.13(a)]. Both Mn^{2+} and Zn^{2+} are able to bind to the soft sulfur ligand, in contrast to Mg^{2+}, which prefers oxygen or nitrogen ligands.[44] Such experiments suggest that Mg^{2+} directly coordinates the phosphate oxygen during catalysis. Magnesium was found to inhibit the reaction of the ribozyme with Mn^{2+} and the thiosubstrate [Fig. 3.13(b)], demonstrating that Mn^{2+} binds at the same site as Mg^{2+}.

3.4.2 Modification of a Substrate to Distinguish Diastereomers of a Metal–Substrate Complex

Many enzymes show a preference for either the Λ or Δ diastereomer of MgATP. For example, hexokinase and acetate kinase show specificity for the Λ isomer, while pyruvate kinase and myosin show specificity for the Δ isomer.[45] Divalent magnesium preferentially binds to oxygen rather than sulfur-derived ligands. Conse-

Table 3.4 KINETIC PARAMETERS FOR RIBOZYME CLEAVAGE OF DNA AND PHOSPHOROTHIOATE DERIVATIVES WITH DIFFERENT DIVALENT METAL IONS

Metal	Metal concentration (mM)	Substrate $5'-^{32}P-$ d(CCCUCU$_x$A)	k_c $(10^{-4}$ min$^{-1})$	K_M or K_D (nM)
Mg^{2+}	12	3'O	8.9	270
		3'S	0.009	100
Mg^{2+}	200	3'O	99	—
		3'S	0.099	—
Mg^{2+}, Mn^{2+}	10,2	3'O	29	~41
		3'S	3.7	~24
Mn^{2+}	12	3'O	21	—
		3'S	6.0	—

Adapted from Ref. 43.

quently, the Mg^{2+} complex with (R_p)-βS-ATP will exist as the Λ complex, and with (S_p)-βS-ATP will exist as the Λ complex.[46]

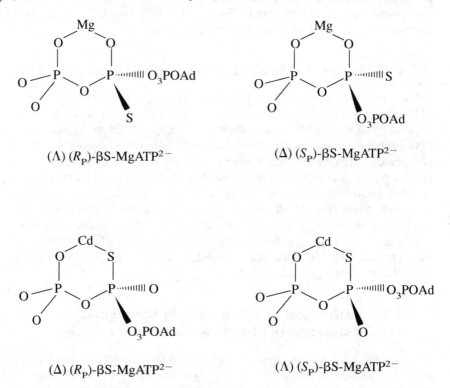

(Λ) (R_p)-βS-MgATP^{2-}

(Δ) (S_p)-βS-MgATP^{2-}

(Δ) (R_p)-βS-MgATP^{2-}

(Λ) (S_p)-βS-MgATP^{2-}

Cd^{2+} preferentially binds to sulfur rather than oxygen, and so (R_p)-βS-ATP forms the Δ complex with Cd^{2+} while (S_p)-βS-ATP forms the Λ complex. This

(a)

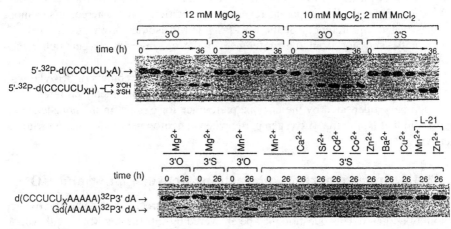

(b)

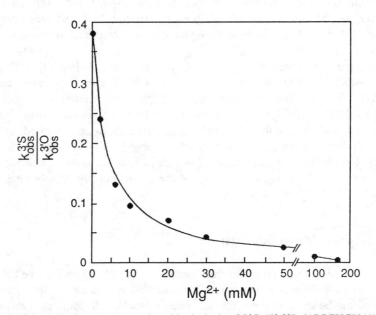

Figure 3.13 (a) Endonuclease-catalyzed hydrolysis of 3'O (5'-^{32}P-d(CCCUCUA)) and 3'S (5'-^{32}P-d(CCCUCU$_S$A)) after 0, 2, 4, 8, 18, and 36 hours. (b) Inhibition of the Mn^{2+}-activated endonuclease-catalyzed hydrolysis of substrate 3'S (5'-^{32}P-d(CCCUCU$_S$A)) by Mg^{2+}, demonstrating that the binding sites for Mg^{2+} and Mn^{2+} are likely to be similar. Adapted from Ref. 43.

links a specific sulfur diastereomer to a chelate isomer. Thus hexokinase, which prefers the Λ chelate, reacts more readily with (R_p)-βS-ATP in the presence of Mg^{2+}, and with (S_p)-βS-ATP in the presence of Cd^{2+}, since in each case the Λ chelate is formed. The apparent reversal of specificity to diastereomers demonstrates that the preference for the (R_p)-βS-ATPMg isomer originates from the Λ specificity, and not from the preference of the enzyme for oxygen rather than sulfur.[40]

In contrast, *E. coli* RNA polymerase utilizes both diastereomers of βS-ATP with Mg^{2+}, but only (R_p)-βS-ATP with Cd^{2+}. Apparently the metal preference for oxygen is compensated by the enzyme preference for oxygen in the nonchelated β position; that is, the free β-oxygen participates in hydrogen bond formation with the enzyme.[47]

3.4.2 Superhyperfine Interactions between Mn^{2+} and ^{17}O

It is often of interest to learn whether Mg^{2+} interacts directly with a specific oxygen atom of a ligand. This question may be answered by substituting Mg^{2+} with Mn^{2+}, and labeling the substrate with ^{17}O at specific positions. ^{17}O has a nuclear spin of $\frac{5}{2}$, which can couple with the unpaired electrons on Mn^{2+} if there is direct binding. This hyperfine coupling may be detected in the EPR spectrum of Mn^{2+}.[48,49] By use of this technique, the stereochemical configuration of the Mn^{2+}–ADP complex at the active site of creatine kinase has been determined.[50] It had previously been determined that the Mn^{2+}-activated enzyme demonstrated 85% of the activity obtained with Mg^{2+}.[49]

These studies required the synthesis of two isomers, R_p [α-^{17}O] ADP and S_p [α-^{17}O] ADP. An equilibrium mixture containing the major species (enzyme· Mn^{2+}ATP·creatine) and (enzyme·Mn^{2+}ADP·phosphocreatine) complexes was studied by EPR. Spectra obtained from this equilibrium mixture, using both unlabeled ADP and each of the two labeled diastereomers, were measured separately (Fig. 3.14). Inhomogeneous broadening was observed with ^{17}O in the pro-S oxygen of the nucleotide, but not with the pro-R oxygen, and so the enzyme selected for the Λ configuration of the α,β-chelate ring of the metal–nucleotide substrates (M^{2+}-ATP and M^{2+}-ADP).[50]

3.5 Enzyme Modification

Site-directed mutagenesis is a powerful method for modifying an enzyme by substitution with natural (or non-natural) amino acids.[51] First, however, the gene of the protein should be cloned, or a synthetic gene constructed. Second, the protein must be expressed in an appropriate host organism.

3.5.1 Identification of Metal-Bound Ligands

It may be possible to identify those amino acid residues that bind a metal cofactor by use of site-directed mutagenesis. Detailed insight on the contributions of specific

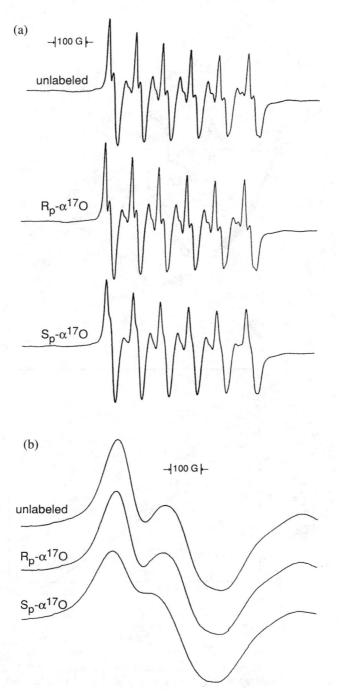

Figure 3.14 EPR spectra obtained for solutions of the equilibrium mixtures of substrates and products with unlabeled nucleotide and α-^{17}O-labeled nucleotides. (a) The spectrum for the central fine-structure transition, and (b) an expansion that shows the lowest-field ^{55}Mn hyperfine component of this set. Adapted from Ref. 50.

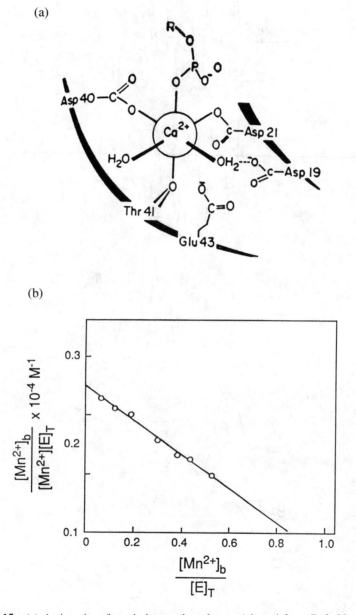

Figure 3.15 (a) Active site of staphylococcal nuclease. Adapted from Ref. 23. (b) Scatchard plot of Mn^{2+} binding to the E43S mutant of staphylococcal nuclease. Adapted from Ref. 52.

residues to the binding free energy of metal cofactors can be obtained by comparing the binding affinity of the native enzyme with that of a mutant protein bearing a residue that cannot ligate the metal ion. For example, staphylococcal nuclease is a Ca^{2+}-dependent nuclease [Fig. 3.15(a)]; however, Mn^{2+} binds at the Ca^{2+} site and may be used as an EPR probe. The role of Glu-43 in metal binding has been studied by use of site-directed mutagenesis by comparing the dissociation constants [obtained by EPR, Fig. 3.15(b) for Mn^{2+} with those obtained for the wild type and an E43S mutant (Glu-43 replaced with serine).[52] The binding constant for the mutant ($K_D = 3400$ μM) was eightfold weaker than that for the wild type ($K_D = 416$ μM). Since the mutation caused only minor changes in the overall structure of the enzyme, as judged by the thermal stability of the enzyme and by NMR spectroscopy, it was concluded that Glu-43 participated in binding the metal cofactor.

In a similar fashion the role of carboxylate residues in the Mg^{2+}-binding pocket of Che-Y, the response regulator of the chemosensory system in *E. coli*, was examined by systematic mutagenesis experiments. With certain mutations the binding affinity for Mg^{2+} dropped by two orders of magnitude.[52] Further examples illustrating the use of site-directed mutagenesis to probe the coordination environments of metal cofactors in important enzymes in nucleic acid biochemistry are described in Chapter 6.

3.6 Closing Remarks

Strategies that use transition metal ions and complexes as spectroscopic and mechanistic probes have been a valuable source of information concerning magnesium biochemistry. With careful choice of metal analogues, and thoughtful consideration of their intrinsic chemistry (both similarities and differences relative to divalent magnesium), information on the structural and catalytic chemistry of essential metal cofactors, and the kinetic and thermodynamic principles that define their mode of action, can be evaluated. While such probe complexes ought generally to mimic the properties of the natural cofactor, it is often the case that an experiment with a metal analogue that differs in a *well-defined* respect to the chemistry of Mg^{2+} may shed most light on its function.

References

1. Cowan, J. A., *Comments Inorg. Chem.* **1992,** *13,* 293–312.

2. Martin, R. B., in *Metal Ions in Biological Systems* (Sigel, H.; Sigel, A., Eds.), Vol. 26, Marcel Dekker, Inc., New York, Basel, 1990.

3. Margerum, D. W.; Cayley, G. R.; Weatherburn, D. C.; Pagenkopf, G. K., in *Coordination Chemistry*, Vol. 2 (Martell, A., Ed.), ACS Monogr. 174, American Chemical Society, Washington, D.C., 1978, p. 1.

4. Cleland, W. W.; Mildvan, A. S., in *Advances in Inorganic Chemistry*, (Eichhorn, G. L.; and Marzilli, L., Eds.), Elsevier North Holland, Inc., 1979, pp. 163–91.

5. Dwek, R. A.; Williams, R. J. P.; Xavier, A. V., in *Metal Ions in Biological Systems* (Sigel, H.; Sigel, A., Eds.), Vol. 4. Marcel Dekker, Inc., New York, Basel, 1974.

6. Martin, R. B., in *Metal Ions in Biological Systems* (Sigel, H.; Sigel, A., Eds.), Vol. 17, Marcel Dekker, Inc., New York, Basel, 1984.

7. Drago, R. S., *Physical Methods in Chemistry*, W. B. Saunders Co., Philadelphia, London, Toronto, 1977.

8. Sanders, J. K. M.; Hunter, B. K., *Modern NMR Spectroscopy*, Oxford University Press, Oxford, New York, Toronto, 1987.

9. Derome, A. E., *Modern NMR Techniques for Chemistry Research* Pergamon Press, Oxford, New York, 1987.

10. Bernheim, R. A.; Brown, T. H.; Gutowsky, H. S.; Woessner, D. E., *J. Chem. Phys.* **1959**, *30*, 950.

11. Gupta, R. K.; Fung, C. H.; Mildvan, A. S., *J. Biol. Chem.* **1976**, *251*, 2421–30.

12. Bean, B. L.; Koren, R.; Mildvan, A. S., *Biochemistry* **1977**, *16*, 3322–33.

13. Stein, P. J.; Mildvan, A. S., *Biochemistry* **1978**, *17*, 2675–84.

14. Gupta, R. K., *J. Biol. Chem.* **1977**, *252*, 5183.

15. Granot, J.; Mildvan, A. S.; Bramson, H. N.; Kaiser, E. T., *Biochemistry* **1980**, *19*, 3537–43.

16. Mildvan, A. S.; Engle, J. L., *Methods Enzymol.* **1972**, *26C*, 654.

17. Serpersu, E. H.; Shortle, D.; Mildvan, A. S., *Biochemistry* **1987**, *26*, 1289–1300.

18. Banci, L.; Bencini, A.; Benelli, C.; Gatteshi, D.; Zanchini, C., *Struct. Bonding* **1982**, *52*, 37–86.

19. Kennedy, F. S.; Hill, H. A. O.; Kaden, T. A.; Vallee, B. L., *Biochem. Biophys. Res. Commun.* **1972**, *48*, 1533–89.

20. Han, H.; Rifkind, J. M.; Mildvan, A. S., *Biochemistry* **1991**, *30*, 11104–8.

21. Mims, W. B.; Peisach, J., in *Biological Magnetic Resonance* (Berliner, L. J.; Reuben, J., Eds)., Vol. III, Plenum, New York, 1981, pp. 213–63.

22. Peisach, J.; Mims, W. B.; Davis, J. L., *J. Biol. Chem.* **1979**, *254*, 12379–89.

23. Serpersu, E. H.; McCracken, J.; Peisach, J.; Mildvan, A. S., *Biochemistry* **1988**, *27*, 8034–44.

24. Tsvetkov, Y. D.; Dikanov, S. A., in *Metal Ions in Biological Systems* (Sigel, H., Ed.), Vol. 22, Marcel Dekker, Inc., New York, Basel, 1987.

25. Mims, W. B.; Peisach, J.; Davis, J. L., *J. Chem. Phys.* **1977**, *66*, 5536–50.

26. Stezowski, J. J.; Hoard, J. L., *Isr. J. Chem.* **1984**, *24*, 323–34.

27. (a) Villafranca, J. J.; Balakrishnan, M. S.; Wedler, F. C., *Biochem. Biophys. Res. Commun.* **1977**, *75*, 464–71. (b) Serpersu, E. H.; McCracken, J.; Peisach, J.; Mildvan, A. S., *Biochemistry* **1988**, *27*, 8034–44.

28. (a) Leigh, J. S., *J. Chem. Phys.* **1970**, *52*, 2608–12. (b) Campbell, I. D.; Dwek, R. A., *Biological Spectroscopy*, The Benjamin/Cummings Publishing Co., Inc., Menlo Park, Reading, London, Amsterdam, 1984.

29. Horrocks, W. DeW., Jr.; Ishley, J. N.; Holmquist, B.; Thompson, J. S., *J. Inorg. Biochem.* **1980**, *12*, 131–41.

30. Bertini, I.; Luchinat, C., *Advances in Inorganic Biochemistry* (Eichhorn, G. L.; Marzilli, L. G., Eds.), Vol. 6, pp. 71–111, Elsevier, New York, 1984.

31. Lehman, I. R.; Richardson, C. C., *J. Biol. Chem.* **1964,** *239,* 233–41; Kornberg, A.; *DNA Replication,* W. H. Freeman, San Francisco, CA, 1980.

32. Holmquist, B.; Vallee, B. L., *Proc. Natl. Acad. Sci. USA* **1979,** *76,* 6216–20.

33. Holmquist, B.; Kaden, T. A.; Vallee, B. L., *Biochemistry* **1975,** *14,* 1454–61.

34. Kim, S.; Cowan, J. A., *J. Inorg. Chem.* **1992,** *31,* 3495–96.

35. Black, C. B.; Cowan, J. A., *J. Am. Chem. Soc.* **1994,** *116,* 1174.

36. Jou, R.; Cowan, J. A., *J. Am. Chem. Soc.* **1991,** *113,* 6685–86.

37. Cornelius, R. D.; Cleland, W. W., *Biochemistry* **1978,** *17,* 3279–86.

38. Cornelius, R. D.; Hart, P. A.; Cleland, W. W., *Inorg. Chem.* **1977,** *16,* 2799.

39. Merritt, E. A.; Sundaralingham, M.; Cornelius, R. D.; Cleland, W. W., *Biochemistry* **1978,** *17,* 3274–78.

40. Dunaway-Mariano, D.; Cleland, W. W., *Biochemistry* **1980,** *19,* 1506–15. (b) Jaffe, E. K.; Cohn, M., *J. Biol. Chem.* **1979,** *254,* 10839–45. (c) Eckstein, F., *Accts. Chem. Res.* **1979,** *12,* 204–10.

41. Cech, T. R., *Science* **1987,** *236,* 1532.

42. Forster, A. C.; Altman, S., *Science* **1990,** *249,* 784.

43. Piccirilli, J. A.; Vyle, J. S.; Caruthers, M. H.; Cech, T. R., *Nature* **1993,** *361,* 85–88.

44. Jaffe, E. K.; Cohn, M., *J. Biol. Chem.* **1978,** *253,* 4823–25.

45. Eckstein, F.; Goody, R. S., *Biochemistry* **1976,** 15, 1685.

46. Jaffe, E. K.; Cohn, M., *Biochemistry* **1978,** *17,* 652.

47. Armstrong, V. W.; Yee, D.; Eckstein, F., *Biochemistry* **1979,** *18,* 4120.

48. Webb, M. R.; Ash, D. E.; Leyh, T. S.; Trentham, D. R.; Reed, G. H., *J. Biol. Chem.* **1982,** *257,* 3068–72.

49. Reed, G. H.; Leyh, T. S., *Biochem.* **1980,** *19,* 5472–80.

50. Leyh, T. S.; Sammons, R. D.; Frey, P. A.; Reed, G. H., *J. Biol. Chem.* **1982,** *257,* 15047–53.

51. Kunkel, T. A., *Proc. Natl. Acad. Sci. USA* **1985,** *82,* 488–92.

52. (a) Bourret, R. B.; Hess, J. F.; Simon, M. I., *Proc. Natl. Acad. Sci. USA* **1990,** *87,* 41–45. (b) Bourret, R. B.; Borkovich, K. A.; Simon, M. I., *Proc. Natl. Acad. Sci. USA* **1987,** *84,* 7609–13.

Modes and Dynamics of Mg^{2+}– Polynucleotide Interactions

Dietmar Porschke

4.1 Introduction

The biological function of polynucleotides is known to be dependent on cooperation of many different components of living systems. Some of these components are usually assumed to be present in the "environment" and often are not mentioned at all. Among these environmental factors are various ions, which are to be present within a given concentration range. The dependence on a certain ionic environment is partly due to the high charge density of polynucleotides,[1,2] which must be reduced by binding of positively charged counterions. This reduction is required, for example, for folding of polynucleotide chains into defined secondary and tertiary structures, which are essential for their biological function. However, in addition to simple compensation of charges, there are more specific effects, indicated by the requirement for special ions. Most biological processes involving polynucleotides require the presence of Mg^{2+}.

The dependence on Mg^{2+} ions is remarkable. It is not immediately apparent why Mg^{2+} cannot be replaced simply by one of several ions, which appear to be quite similar. Ca^{2+} ions, for example are very closely related to Mg^{2+}, but there is an explicit antagonism of Mg^{2+} and Ca^{2+} in biochemical reactions.[3,4] In the present contribution I will attempt to explain the special role of Mg^{2+} on the basis of its dynamics. The dynamics of Mg^{2+} complexation is known to be clearly different from that of other ions.[5] First, this difference is described on the basis of data obtained for various "simple" ligands. Then it will be shown how the difference is reflected in the interactions of Mg^{2+} with polynucleotides at various levels of secondary and tertiary structure. The available data clearly demonstrate special modes of interactions between polynucleotides and Mg^{2+} ions, which are very probably important for biological function.

4.2 A Short Comparison of Mg^{2+} with Other Ions: Thermodynamics and Kinetics of Complex Formation

Complexes of a large number of ligands with most of the metal ions of the periodic system have been investigated with respect to both thermodynamics and kinetics.[5-7] Owing to these investigations the most important parameters of complex formation have been characterized. As an introduction to the subject of the present contribution, it is useful to present a short summary of the results—with an emphasis on alkali and alkaline earth metal ions.

It is important to recall that the free energies of hydration,[6,7] are very high (cf. Table 4.1). Obviously, replacement of water molecules around metal ions requires ligands with at least similar energies of interaction. Some ligands will not be able to replace the water molecules, which are directly attached to the metal ion. In this case, complex formation will be restricted to contacts via the "inner" hydration layer by electrostatic interactions. The high energies of hydration may also lead to high activation barriers, which have to be passed, when water is replaced by other ligands.

Equilibrium constants for complex formation by metal ions have been measured for many different ligands in aqueous solution.[8,9] A comparison of some values compiled in Table 4.2 shows that for a given ligand the binding constants are clearly larger for the alkaline earth metal ions than for the alkali metal ions. This is mainly due to the higher charge of the alkaline earth ions. Within the series of alkali or of alkaline earth metal ions, the binding constants for a given ligand usually show very

Table 4.1 ENTHALPY AND FREE ENERGY OF HYDRATION OF ALKALI METAL AND ALKALINE EARTH METAL IONS (kJ/mol; 25°C; from Ref. 7)

Ion	$-\Delta H$	$-\Delta G$
Li^+	515	511
Na^+	406	411
K^+	321	337
Rb^+	296	316
Cs^+	263	284
Be^{2+}	2488	2443
Mg^{2+}	1923	1907
Ca^{2+}	1593	1594
Sr^{2+}	1445	1448
Ba^{2+}	1304	1319

Table 4.2 STABILITY CONSTANTS OF METAL ION COMPLEXES (LOG K_1 IN AQUEOUS SOLUTION)

Ligand	Glutamic acid	HPO_4^{2-}	ATP	EDTA
Conditions	25°C, 0.1 M	25°C, 0.5 M	25°C, 0.1 M	20°C, 0.1 M
Ref.	8	8	9	8
Li^+	—	0.72	1.7	2.79
Na^+	—	0.60	0.98	1.66
K^+	—	0.49	0.99	0.8
Be^{2+}	—	—	—	9.2
Mg^{2+}	1.9	1.8	4.4	8.79
Ca^{2+}	1.43	1.6	3.92	10.69
Sr^{2+}	1.37	1.4	3.60	8.73
Ba^{2+}	1.28	—	3.73	7.86

little variation. Some simple examples for the absence of large variations of the binding constant within a given class of metal ions will be presented for the case of oligo- and polynucleotides. Thus it is not possible to explain the requirement for Mg^{2+} ions in the nucleic acid metabolism and the antagonism of Ca^{2+}/Mg^{2+} on the basis of equilibrium parameters. It should be mentioned, however, that there are special ligands with binding pockets of exactly adapted geometry that are able to distinguish Mg^{2+} and Ca^{2+} ions on the basis of their binding constants.

The rates of complex formation between metal ions and various ligands are very high in many cases and, thus, investigation of the kinetics requires fast reaction techniques.[10,11] Extensive studies of many metal ion complexes by these techniques have shown that the metal ions may be classified into three different groups.[5] In a first group the water molecules in the "inner" hydration sphere of the metal ions are relatively labile and may be replaced by other ligands without any perceptible activation barrier. In this case the rate of complex formation is determined by the rate of diffusion, and it is not possible to separate complexes with ligands in the "outer" hydration sphere and in the "inner" hydration sphere. Examples of this type of ion are the alkali metal ions.

A second group of metal ions represents the other extreme, with very strong contacts between water molecules and the metal ion. In this case, dissociation of whole H_2O molecules from the inner hydration sphere is associated with particularly high activation barriers, and, thus, substitution does not proceed by dissociation of H_2O units, but by stepwise dissociation of H^+ or of OH^-, that is, by hydrolysis. The overall rate is dependent on the nature of the ligand and its basicity. Examples for this group are highly charged ions with a small ion radius like Be^{2+}, Al^{3+}, and Fe^{3+}.

The third and largest group of metal ions shows a characteristic rate for ligand substitution, which is almost independent of the nature of the ligand. Complexes are formed in a two-step reaction: In a first step the ligand enters into the outer hydration sphere and forms an "outer-sphere" complex; in a second step the ligand may replace one or more of the water molecules attached directly to the metal ion and may form an "inner-sphere" complex.

$$\text{metal} + \text{ligand} \rightleftharpoons \text{outer-sphere complex} \rightleftharpoons \text{inner-sphere complex} \qquad (4.1)$$

In this group of metal ions the rate of substitution of water molecules in the inner hydration sphere is mainly dependent on the metal ion and almost independent of the ligand. A typical member of this group is Mg^{2+}.

As shown by the data compiled in Figure 4.1, Mg^{2+}–inner-sphere complexes are formed with a rate constant, which is by orders of magnitudes different from those of its immediate neighbors in the periodic system Ca^{2+} and Be^{2+}. Ca^{2+} forms its inner-sphere complexes with a very high rate constant of about 3×10^8 s^{-1}, and thus the kinetics of complex formation by Ca^{2+} ions is rather similar to that observed for alkali metal ions. The rate constant for inner-sphere complexation by Mg^{2+} ions is around 2×10^5 s^{-1}, which is more than three orders of magnitude smaller than that observed for Ca^{2+}. Finally, the characteristic rate constant for inner-sphere complexation of Be^{2+} ions, $\approx 3 \times 10^3$, is again three orders of magnitude smaller than that found for Mg^{2+}. These differences have been observed

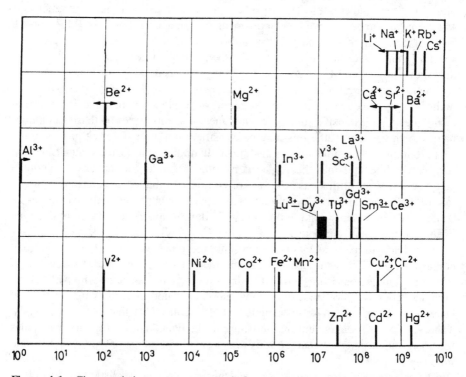

Figure 4.1   Characteristic rate constants (s^{-1}) for substitution of inner-sphere H_2O of various ions in aqueous solution according to Diebler et al.[5] (reprinted by permission of IUPAC)

not only for the replacement of H_2O molecules by other ligands, but also for the replacement of H_2O by H_2O measured by NMR techniques.[12] Thus the differences in the rate constants of inner-sphere complexation clearly reflect different activation barriers in the dissociation of H_2O from the inner hydration sphere. An illustration of the substitution process for the case of Mg^{2+}, which is surrounded by six water molecules in its hydrated state, is shown in Figure 4.2. As will be shown, the characteristic complexation rate constants are also valid for the formation of complexes with oligo- or polynucleotides and can be used for quantitative characterization of these complexes.

4.3 Relaxation Methods: Advantages and Limitations

Most of the results compiled in the present contribution have been obtained by relaxation methods, which have been described elsewhere.[10,11] The technical potential provided by these methods has been used for the characterization of various types of metal ion interactions with polynucleotides. However, there are still some gaps in the knowledge on the dynamics of ion interactions with polynucleotides, which are due to some technical limits. Thus the selection of systems to be described should be explained by a short introduction into the technical potential of the available methods.

Chemical relaxation is induced by rapid perturbation of a chemical equilibrium. The relaxation towards the new equilibrium is usually recorded by measurement of an appropriate spectroscopic parameter. Both perturbation and observation must be sufficiently fast compared to the reaction time constant; for a sufficiently large

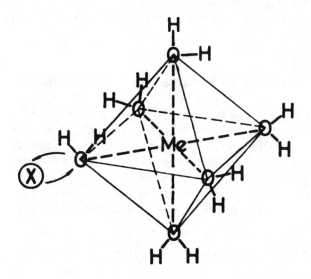

Figure 4.2   Metal ion with six water molecules in an octahedral coordination.

change of the observation parameter upon perturbation, the modes of perturbation and observation have to be adapted to the system under investigation. Although this adaptation is possible in most cases, some problems remain. For example, Diebler et al.[13] did not find any measurable amplitude in their investigation of Mg^{2+} interactions with oligoriboadenylates by the temperature jump technique. This was attributed to a dominance of electrostatic effects in the interactions, which implies enthalpy changes close to zero and, thus, virtually no perturbation by temperature jumps. Diebler and co-workers[14,15] found sufficient amplitudes for the binding of various transition metal ions to nucleic acids, such as Ni^{2+} and Co^{2+}, which do interact not only by electrostatics but also by coordination to base residues.

The equilibrium of complexes with a large contribution from electrostatic interactions may be perturbed by electric field pulses.[16] As will be described, the field pulse method has been particularly useful in the characterization of Mg^{2+} complexes with nucleic acids. However, electric field pulses applied to polynucleotides do not only induce perturbation of ion binding but also orientation of the polynucleotide chains into the direction of the field vector, which is often associated with considerable changes in the absorbance ("electric dichroism"). Separation of "chemical" reaction effects from "physical" orientation effects requires a special technique: Chemical relaxation in the presence of orientation effects may be recorded selectively, when the absorbance of the system is measured by linearly polarized light with the plane of polarization oriented at the "magic" angle of 54.8° with respect to the field vector.[17-19] However, successful application of this technique requires careful adjustment of the optical detection setup. Some reports in the literature demonstrate the technical problems,[18,20] which may be associated with experiments based on the magic angle technique. As shown by many applications in the laboratory of the author, these problems can be solved by careful preparation of the experiments. The magic angle technique is useful, when the magnitude of the optical effect resulting from chemical relaxation is not much lower than that resulting from orientation effects.

4.4 Mg^{2+} Binding to Single-Stranded Polynucleotides

Binding of ions to single-stranded polynucleotides is accompanied by an increase of stacking interactions between adjacent base residues.[21] This effect is very useful for detection of ion binding, because binding of ions like Na^+ and Mg^{2+} itself is not associated with any sufficiently large change of the UV absorbance of polynucleotides. Changes in the stacking interactions induced by ion binding lead to rather large absorbance changes, which can be used to measure the coupled relaxation process. The coupled reaction may be described by the simplified scheme

$$\text{ion} + \text{polymer} \cdot \text{S1} \underset{k_1^-}{\overset{k_1^+}{\rightleftharpoons}} \text{ion} \cdot \text{polymer} \cdot \text{S1} \underset{k_2^-}{\overset{k_2^+}{\rightleftharpoons}} \text{ion} \cdot \text{polymer} \cdot \text{S2} \tag{4.2}$$

where polymer·S1 denotes the polymer with a low degree of base stacking, ion·polymer·S1 the ion–polymer complex with a low degree of stacking, and ion·polymer·S2 the ion–polymer complex with a higher degree of stacking. The coupled equilibrium has been perturbed by electric field pulses,[21] which may induce very large changes of the ion binding equilibrium. Nevertheless, the simple linearized relaxation equations may be used for the data evaluation, because under standard conditions the bimolecular step is a quasi-first-order reaction due to a large excess of the ion concentration over the polymer concentration.

Investigations of various single-stranded polynucleotides by the electric field jump technique in the presence of *monovalent* ions showed that binding of these ions is very fast compared to the stacking reaction. In this case, the relaxation time constants measured at different ion concentrations provide the rate constants of the stacking reaction k_2^+ and k_2^- together with the apparent binding constant of the ions to the polymer $K_1 = k_1^+/k_1^-$, but not the rate constants of ion binding.[21]

More information has been obtained for the reaction of Mg^{2+} and Ca^{2+} with single-stranded polynucleotides. In this case, the effective binding constant is much higher than that for monovalent ions, and, thus, the rate constant of dissociation is clearly lower than that for monovalent ions. Due to this fact, a separate relaxation process reflects binding of Mg^{2+} or of Ca^{2+} ions[21] with a time constant clearly larger than that observed for the stacking reaction coupled with binding of monovalent ions (cf. Fig 4.3).

The relaxation process reflecting binding of Mg^{2+} and of Ca^{2+} is very useful for

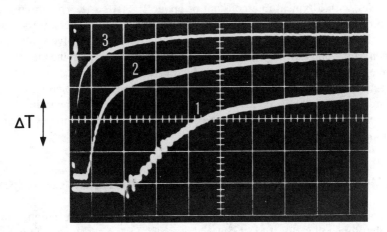

Figure 4.3 Electric field jump relaxation measured at 248 nm (magic angle) for poly(C) (58 µM monomer units) in a buffer containing 10 µM MgCl$_2$ and 0.5 mM Tris, pH 8.0, after a short pulse of 53.7 kV/cm. The relaxation signal has been recorded at three different time scales: (1) 0.5 µs, (2) 2 µs, and (3) 10 µs (each per large scale unit). The amplitude units (ΔT) are 2% change of transmission per large scale unit (starting points shifted for each shot; from Ref. 21).

characterization of the state of ion binding. Measurements of the time constant τ_2 associated with this process at various ion concentrations $[Me^{2+}]$ show the existence of a range of low $[Me^{2+}]$ values, where τ_2 is almost independent of $[Me^{2+}]$. Above a limiting $[Me^{2+}]$ value, a strong decrease of τ_2 is observed. This result is readily explained by the stoichiometry of the binding reaction and by the very high binding affinity. When Mg^{2+} ions are added to the polymer, these ions are almost completely bound to the polymer as long as the binding capacity of the polymer is not exceeded. In this range τ_2 is almost independent of the total $[Me^{2+}]$ concentration. When the binding capacity is exceeded, the free concentration $[Me_f^{2+}]$ increases strongly with the total concentration $[Me^{2+}]$, and thus the relaxation time constant decreases as expected for a bimolecular reaction. The experimental data show that the stoichiometry is not one bivalent metal ion per two phosphates, but only 0.4 to 0.6 metal ions per two phosphates.

Obviously, the relaxation time constants also provide information on the kinetics of the reaction. The experimental data[21] show that both Mg^{2+} and Ca^{2+} ions bind to poly(A) and to poly(C) with a rate constant of 1 to 2×10^{10} M^{-1} s^{-1}. These rate constants indicate reactions that are controlled by diffusion. Each encounter of the reaction partner leads to complex formation. This result is expected for simple types of bimolecular reactions, when the reaction pathway is without any large activation barriers.

A comparison with the kinetics of various Mg^{2+}-complexation reactions reveals that there is something special about the complexes of Mg^{2+} ions with polynucleotides. For all cases of strong Mg^{2+} complexes with various ligands, a slow relaxation process has been observed, which reflects formation of inner-sphere complexes with a typical rate constant of about 2×10^5 s^{-1}. The absence of this slow process and the high rate of complex formation clearly demonstrate that the Mg^{2+} complex formed by poly(A) and poly(C) does not involve direct contacts between Mg^{2+} and any polymer residue. The water molecules in the inner hydration sphere of Mg^{2+} remain unaffected upon binding to single-stranded polynucleotides. As discussed in section 4.8, the preference for ion atmosphere binding of Mg^{2+} to standard stacked polynucleotides appears to be due mainly to the rather large distance between subsequent phosphate residues, which prevents optimal charge compensation by site binding.

4.5 Mg^{2+} Binding to Single-Stranded Oligonucleotides

During some simple test experiments by the electric field jump technique, an unexpectedly slow relaxation process was detected[22] for an aqueous solution of oligoadenylates in the presence of Mg^{2+} ions. The relaxation effect appeared with a conveniently detectable amplitude and a time constant around 100 μs (cf. Fig. 4.4). When Mg^{2+} was replaced by Ca^{2+}, the solutions did not show any slow relaxation effect. This result clearly indicates the formation of an inner-sphere complex by Mg^{2+} ions. Controls by various methods showed that addition of Mg^{2+} or Ca^{2+} did not induce any aggregation of the oligonucleotide. Thus the relaxation effect clearly

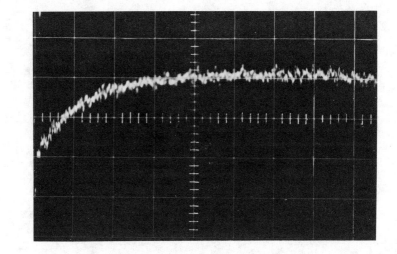

Figure 4.4   Electric field jump relaxation of a solution containing 16.7 μM A(pA)₅ and 100 μM MgCl₂ observed after an electric field pulse of 30 kV/cm for 100 μs (0.5 mM Tris, pH 8.0, 20°C; from Ref. 22).

reflects binding of Mg^{2+} to the single-stranded oligomer without complications by any "secondary" reactions.

Starting from the unequivocal assignment of the relaxation effect to Mg^{2+}–inner-sphere complexation, the effect has been used for a quantitative characterization of Mg^{2+}–inner-sphere complexes formed with oligoriboadenylates.[22] The inner-sphere complexes CI are formed via outer-sphere complexes CO

$$Mg^{2+} + \text{oligo(A)} \overset{k_1^+}{\underset{k_1^-}{\rightleftharpoons}} CO \overset{k_2^+}{\underset{k_2^-}{\rightleftharpoons}} CI \qquad (4.3)$$

According to the theory,[10,11] the mechanism should be reflected by two relaxation processes, but the experiments do not reveal both processes in all cases. The first reaction step resulting in the formation of outer-sphere complexes is much faster than inner-sphere complexation and is well separated on the time scale from the second step. Furthermore, the first step is associated with a rather small amplitude for the short oligomers, and thus the corresponding relaxation process cannot be detected. However, the short oligomers exhibit particularly large amplitudes associated with the second relaxation process. As shown in Figure 4.5 for the case of A(pA)₄, the reciprocal time constants $1/\tau_2$ associated with this process increase

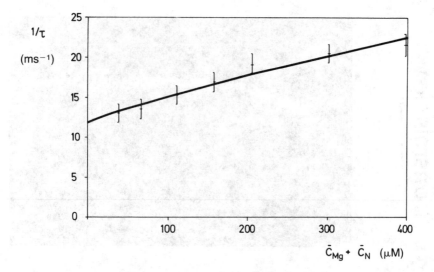

Figure 4.5 Reciprocal relaxation time $1/\tau$ as a function of the free reactant concentration $\bar{c}_{Mg} + \bar{c}_N$ for $A(pA)_5 + Mg^{2+}$ in 0.5 mM Tris, pH 8.0. The solid line represents a least-squares fit according to Eq. (4.4) (from Ref. 22).

almost linearly with the Mg^{2+} concentration. The concentration dependence of $1/\tau_2$ together with the overall binding constant $K = (k_1^+/k_1^-)(1 + k_2^+/k_2^-)$, which has been determined by spectrophotometric titrations, have been used to evaluate the rate constants k_2^+ and k_2^- together with the equilibrium constant k_1^+/k_1^-. This evaluation was based on the general equations for the time constants τ_1 and τ_2 of the two relaxation processes:

$$\frac{1}{\tau_1} = \frac{1}{2}\{\Sigma k + [(\Sigma k)^2 - 4 \sqcap k]^{1/2}\} \tag{4.4}$$

$$\frac{1}{\tau_2} = \frac{1}{2}\{\Sigma k - [(\Sigma k)^2 - 4 \sqcap k]^{1/2}\} \tag{4.5}$$

where

$$\Sigma k = k_1^+ (\bar{c}_{Mg} + \bar{c}_N) + k_1^- + k_2^+ + k_2^-$$
$$\sqcap k = k_1^+ (\bar{c}_{Mg} + \bar{c}_N)(k_2^+ + k_2^-) + k_1^- k_2^-$$

c_{Mg} and c_N denote the concentrations of Mg^{2+} ions and of oligonucleotides, respectively. Both relaxation processes have been observed for the oligomer $A(pA)_{17}$ and have been characterized at different concentrations (cf. Fig. 4.6). In this case all rate constants have been determined by a least-squares fit of the relaxation time constants according to Eqs. (4.4) and (4.5)—again using the overall binding constant obtained from equilibrium titrations:

$$k_1^+ = 3.5 \times 10^{10} \text{ M}^{-1} \text{ s}^{-1}$$
$$k_1^- = 1.9 \times 10^5 \text{ s}^{-1}$$
$$k_2^+ = 1.0 \times 10^3 \text{ s}^{-1}$$
$$k_2^- = 2.7 \times 10^4 \text{ s}^{-1}$$

A compilation of the various parameters obtained for different oligomers is given in Table 4.3. The rate constant for formation of inner-sphere complexes k_2^+ decreases with increasing chain length, whereas the rate constant for dissociation of inner-sphere complexes k_2^- increases with increasing chain length. These dependences of the rate constants combine to a very clear decrease of the inner-sphere binding constant $K_2 = k_2^+/k_2^-$ with increasing chain length.

Relaxation measurements by the electric field jump technique have been extended to other oligonucleotides,[22] including C(pC)$_5$, I(pI)$_5$, U(pU)$_5$, and d[A(pA)$_5$]. None of these oligonucleotides showed effects similar to the ones observed with oligoriboadenylates. The changes of the UV absorbance and of the CD spectrum upon addition of Mg^{2+} to these oligonucleotides are very small. Application of electric field pulses did not induce any detectable slow relaxation effect corresponding to inner-sphere complexation. Although it cannot be excluded that inner-sphere complexation by these oligomers has not been detected because of too-low changes of the spectroscopic parameters used in the experiments, it is more

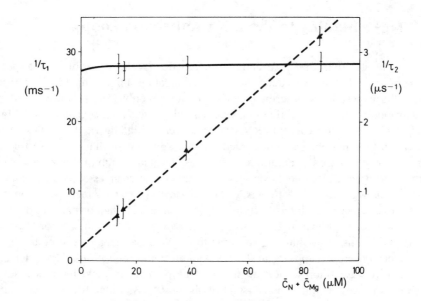

Figure 4.6 Reciprocal relaxation times $1/\tau_1$ (·) and $1/\tau_2$ (Δ) as a function of the free reactant concentration $\bar{c}_{Mg} + \bar{c}_N$ for A(pA)$_{17}$ + Mg^{2+} in 0.5 mM Tris, pH 8.0, at 20°C. The lines represent a least-squares fit according to Eqs. (4.4) and (4.5) (from Ref. 22).

Table 4.3   INCREASE OF CD AMPLITUDE AT 265 nm ΔCD (IN %)
UPON FORMATION OF A $(pA)_n$–Mg^2 COMPLEXES, INNER-SPHERE
BINDING CONSTANTS K_2 *AS WELL AS RATE OF CONSTANTS*
FOR INNER-SPHERE FORMATION k_{12} AND DISSOCIATION k_{21}
(1 mM Na^+ CACODYLATE, pH 5.9, AND 0.5 mM TRIS-HCl, pH 8;
FROM REF. 22)

	pH	ΔCD	K_2	k_{12} (s^{-1})	k_{21} (s^{-1})
A $(pA)_4$	5.9	74.9	4.3	4.9×10^4	1.1×10^4
A $(pA)_5$	5.9	72.5	8.0	7.0×10^4	0.88×10^4
A $(pA)_9$	5.9	40.2	0.2	0.4×10^4	2.0×10^4
A $(pA)_4$	8.0	77.5	5.4	6.5×10^4	1.2×10^4
A $(pA)_5$	8.0	62.6	5.0	4.9×10^4	0.99×10^4
A $(pA)_6$	8.0	49.7	3.8	5.0×10^4	1.3×10^4
A $(pA)_7$	8.0	36.6	0.56	1.0×10^4	1.8×10^4
A $(pA)_9$	8.0	28.5	0.27	0.54×10^4	2.0×10^4
A $(pA)_{17}$	8.9	13.4	0.04	0.1×10^4	2.7×10^4
poly (A)	8.0	10.9	—	—	—

likely that these oligomers do not form inner-sphere complexes as much as oligoriboadenylates.

4.6 Mg^{2+} Binding to the tRNA Anticodon Loop

During investigations of tRNAs and their interactions with various ligands, a special mode of Mg^{2+} binding was detected by the temperature-jump technique.[23] In analogy to the experiments described in the previous section for the case of simple oligonucleotides, the investigation was started by a simple exchange of Mg^{2+} against Ca^{2+}. In the case of tRNA[Phe] from yeast, binding of these ions to the anticodon loop may be easily recorded by measurements of the Wye base fluorescence. The Wye base is a natural modified base located next to the anticodon triplet of tRNA[Phe] (yeast); its fluorescence is very sensitive to changes of its environment. Addition of Mg^{2+} and Ca^{2+} to tRNA[Phe] (yeast) leads to a strong increase of the Wye base fluorescence, which has been used to determine binding constants at different temperatures. As shown in Table 4.4, the thermodynamic parameters for the binding of these ions to the binding site next to the Wye base are very similar.

A clear difference appears in the temperature-jump relaxation—again detected by the Wye base fluorescence. In the absence of bivalent ions, there is a single relaxation process with a time constant of about 100 μs, which has been attributed to a transition between two conformations of the anticodon loop.[23,24] It is likely that these conformations represent the opposite stacking forms of the anticodon triplet towards the 3' and to the 5' ends of the nucleotide chain.[25] Addition of Ca^{2+} leads to changes of both the relaxation time and of the relaxation amplitude of this relaxation process, whereas addition of Mg^{2+} leads to the appearance of a second

Table 4.4   THERMODYNAMIC PARAMETERS FOR THE BINDING
OF Mg²⁺ AND Ca²⁺ TO tRNA^Phe (YEAST) FROM
FLUORESCENCE TITRATIONS[a] (FROM REF. 23)

	pH 6.0		pH 7.1	
	Mg²⁺	Ca²⁺	Mg²⁺	Ca²⁺
ΔH (kcal/mol)	−2.5	−2.4	+0.7	−1.8
ΔS (eu)	4.9	6.3	17.6	9.6
$K(3.4°C)$	1210	1820	1930	3520
$\Delta(3.4°C)$	2.75	2.17	1.874	1.392

[a] Estimated accuracy ±10%.

relaxation process[23] with a time constant of a few milliseconds (cf. Fig. 4.7). Because the binding constants of Ca²⁺ and Mg²⁺ are virtually identical, the appearance of a separate relaxation process in the case of Mg²⁺ clearly indicates the formation of an inner-sphere complex.

The details of the complex formation and its dynamics have been determined by measurements of relaxation time constants and amplitudes over a broad range of

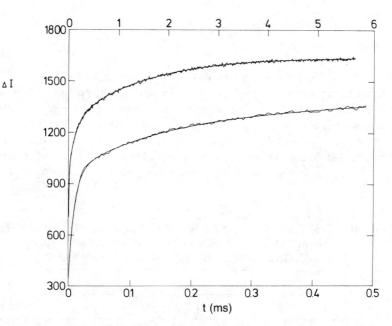

Figure 4.7   Temperature jump relaxation of tRNA^Phe(yeast) detected by measurement of the Wye base fluorescence. The relaxation effect is shown at two different times scales—the upper and lower time scales are for the upper and lower curves, respectively. There are two relaxation effects with time constants $\tau_1 = 129$ μs and $\tau_2 = 1.45$ ms in addition to the standard fast thermal quenching (3.4°C; 0.1 M NaClO₄, 0.05 M Tris-cacodylate, pH 6.0, 0.3 mM Mg (ClO₄)₂ (from Ref. 23).

concentrations. It has been shown that the most simple mechanism consistent with the data[23] obtained in the presence of Ca^{2+} is as follows:

$$A \underset{k_A^-}{\overset{k_A^+}{\rightleftharpoons}} B \tag{4.6a}$$

$$B + Ca \underset{k_B^-}{\overset{k_B^+}{\rightleftharpoons}} C \tag{4.6b}$$

where A and B denote the two conformations of the anticodon loop. Ca^{2+} binds selectively to the conformation B and forms the complex C. The binding and rate constants evaluated for this mechanism are compiled in Table 4.5.

The simplest mechanism consistent with the data[23] obtained for the case of Mg^{2+} (cf. Fig. 4.8) is given by

$$A \underset{k_A^-}{\overset{k_A^+}{\rightleftharpoons}} B \tag{4.7a}$$

$$B + Mg \underset{k_o^-}{\overset{k_o^+}{\rightleftharpoons}} CO \underset{k_i^-}{\overset{k_i^+}{\rightleftharpoons}} CI \tag{4.7b}$$

Again A and B denote the two conformations of the anticodon loop; Mg^{2+} binds selectively to the conformation B and forms an outer-sphere complex CO in the first step followed by formation of an inner-sphere complex CI in a second step. The parameters evaluated according to this mechanism are compiled in Table 4.6.

It should be noted that the separation of two reactions steps associated with Mg^{2+} binding and the determination of the parameters for Mg^{2+} inner-sphere complexation is enabled by the low rate of H_2O substitution characteristic of Mg^{2+}. It is very

Table 4.5 THERMODYNAMIC, KINETIC, AND FLUORESCENCE PARAMETERS OBTAINED FOR THE BINDING OF Ca^{2+} TO tRNA[Phe] FROM RELAXATION TIMES AND AMPLITUDES[a] (FROM REF. 23)

	pH 7.1		pH 6.0	
	3.4°C	7.2°C	3.4°C	7.2°C
K (M^{-1})	3540	3350	1820	1720
K_A	0.9	1.0	1.0	0.9
K_B (M^{-1})	7300	6600	3600	3600
k_A^+ (s^{-1})	5.0×10^3	6.4×10^3	3.8×10^3	5.1×10^3

[a] Estimated accuracy $\pm 10\%$; for K_A, $\pm 25\%$.

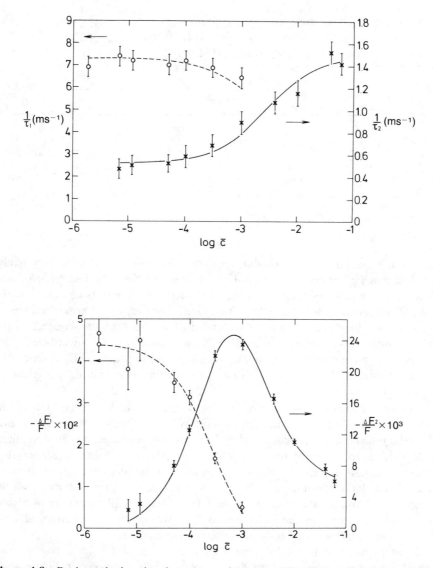

Figure 4.8 Reciprocal relaxation times $1/\tau_1$ and $1/\tau_2$ together with the relative fluorescence amplitudes $\Delta F_1/F$ and $\Delta F_2/F$ as a function of the free Mg^{2+} concentration (logarithmic scale) at 3.4°C in 0.1 M NaClO₄, 0.05 M Tris-cacodylate, pH 6.0. The lines represent least-squares fits with the parameters given in Table 4.6 (from Ref. 23; additional parameters used in the fit, cf. Ref. 23).

likely that binding of Ca^{2+} ions to the anticodon loop involves outer- and inner-sphere complexes as well, but the individual reaction steps cannot be separated because of the high rate of substitution into the inner hydration sphere of Ca^{2+}.

tRNA molecules are rather large, and thus it may be argued that some of the relaxation effects described may be due to coupling of the anticodon loop with other

Table 4.6 THERMODYNAMIC AND KINETIC PARAMETERS OBTAINED
FOR THE BINDING OF Mg^{2+} TO tRNA FROM RELAXATION
TIMES AND AMPLITUDES[a] (FROM REF. 23)

	pH 7.1		pH 6.0	
	3.4°C	7.2°C	3.4°C	7.2°C
K (M^{-1})	2000	2000	1200	1140
K_A	0.6	0.9	1.2	0.4
K_o (M^{-1})	1460	1240	780	1190
K_i	3	2.5	1.9	2.4
k_A^+ (s^{-1})	3.4×10^3	5.9×10^3	4.4×10^3	3.0×10^3
k_i^+ (s^{-1})	1.4×10^3	1.9×10^3	1.0×10^3	1.9×10^3

[a] Estimated accuracy $\pm 10\%$; for K_A and K_i, $\pm 25\%$.

parts of the macromolecule. This interpretation does not appear to be very likely, but it cannot be completely excluded on the basis of the experiments described. However, a remote coupling effect may be ruled out by results obtained for the isolated anticodon loop[26] of tRNA[Phe]. This loop comprises 15 nucleotide residues of the anticodon loop and the anticodon stem of tRNA[Phe]. Thus the sequence of the isolated anticodon loop is less than 20% of the original tRNA. Nevertheless, this fragment shows almost the same relaxation effects as the complete tRNA molecule. A comparison of the parameters evaluated from the relaxation data is given in Figure 4.9.

Further reduction of the sequence to a hexamer, which is excised from the anticodon loop of tRNA[Phe] and comprises the Wye base, leads to the disappearance of all the relaxation effects described.[26] This hexamer no longer has enough nucleotide residues for the formation of a double-helical stem, and thus the special loop structure cannot be formed. We may conclude that the loop structure is essential both for the loop transition and for inner-sphere complexation. Although all the nucleotides with the attachment sites for inner-sphere complexation are available, the inner-sphere complex cannot be formed, because the sites are not arranged in the special orientation required for inner-sphere complexation.

The structure of the Mg^{2+}–inner-sphere complex in the anticodon loop of tRNA[Phe] has been determined by X-ray analysis.[27–31] As shown in Figure 4.10, there is one direct contact of the metal with the phosphate of residue 37—one water molecule of the Mg^{2+} inner hydration sphere is replaced by the oxygen of the phosphate residue. There are many other contacts, but all these other contacts are via water molecules of the Mg^{2+} inner hydration sphere. One of these H_2O-mediated contacts is to the Wye base and provides an explanation for the strong change of the Wye base fluorescence induced by Mg^{2+} binding. Other H_2O-mediated contacts are formed with different residues of the loop. These contacts constitute a network that greatly restricts the mobility of the nucleotide residues. The network (Fig. 4.10) also implies selectivity of Mg^{2+} binding to one conformation—in agreement with the results of the relaxation measurements. In the

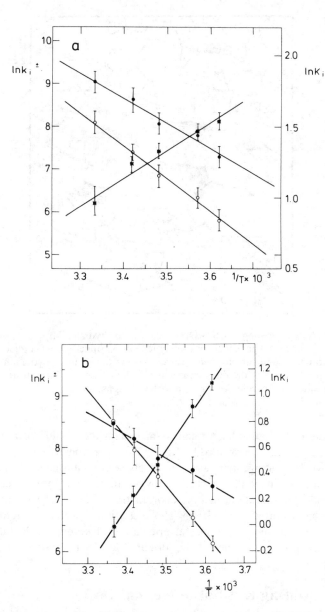

Figure 4.9 Parameters of the Mg^{2+} inner-sphere complexation with (a) the pentadecamer and (b) the complete tRNA^Phe(yeast) as a function of the reciprocal absolute temperature k_i^+ (•), k_i^- (○), $K_i = k_i^+/k_i^-$ (*) (20°C, 100 mM $NaClO_4$, 50 mM Tris-cacodylate, pH 7.1). According to these data, Mg^{2+}–inner-sphere complexation to the pentadecamer is associated with an enthalpy change $\Delta H_i = -15$ kJ/mol, an entropy change $\Delta S_i = -43$ J mol⁻¹ deg⁻¹ and an Arrhenius activation enthalpy $\Delta E_i^+ = +49$ kJ/mol; the corresponding parameters for the whole tRNA^Phe are $\Delta H_i = -36$ kJ/mol, $\Delta S_i = -124$ J mol⁻¹ deg⁻¹ and $\Delta E_i^+ = +43$ kJ/mol (from Ref. 26).

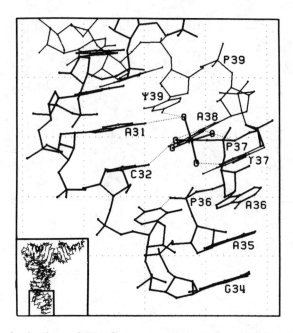

Figure 4.10 Anticodon loop of tRNA^{Phe}(yeast) with the hydrated Mg^{2+} ion. The Mg^{2+} ion is directly coordinated to a phosphate oxygen of residue 37 (inner-sphere complex). Extensive additional hydrogen bonds are formed with the waters of the inner hydration sphere. The insert shows the location of the diagram relative to the whole molecule (from Teeter et al., Ref. 31, reproduced by permission of John Wiley & Sons, Ltd).

crystal structure the anticodon triplet is stacked to the 3′ side of the nucleotide chain, and thus the form *B* defined by Eq. (4.7a) corresponds to the 3′ stack. In the absence of Mg^{2+} binding, there are more degrees of conformational freedom, and the anticodon triplet may also be stacked toward the 5′ end [form *A* defined in Eq. (4.7a)]. Quigley et al.[29] described the coordination of Mg^{2+} in the anticodon loop as a "somewhat distorted octahedron." This distortion as well as the rather complex coordination, which requires a special orientation of several nucleotide residues, explains the relatively low rate for the formation of the inner-sphere complex.

4.7 Na$^+$ Binding to Double-Helical DNA

Unfortunately, it has not been possible yet to analyze the dynamics of binding of alkali or alkaline earth ions to double-helical DNA by the "simple" procedures described, which have been applied successfully for different RNA molecules. A major difficulty associated with double-helical DNA is the absence of sufficiently large spectroscopic signal changes upon ion binding. DNA double helices are relatively rigid, and thus changes in the ion cloud surrounding DNA are hardly transmitted into changes of the base stacking geometry or of the degree of base stacking, which would result in some change of the UV absorbance.

Because of this difficulty, some other procedure has to be used for the characterization of the ion dynamics around double helices. A special approach to the problem was found during measurements of the electric dichroism of double helices[32]: Electric field pulses are used for (partial) alignment of double helices into the direction of the field vector; the alignment is recorded by measurements of the linear dichroism. Linear DNA molecules are virtually symmetric with respect to their charge distribution and thus do not have any large electric dipole moment in the absence of an electric field. Field-induced alignment requires generation of a dipole—in the case of DNA double helices the dipole is generated by some change of the ion binding. Thus it is expected that the alignment of DNA double helices induced by electric field pulses is preceded by generation of a dipole, which is equivalent to a change of ion binding. It must be expected that these two processes appear as a convolution product.

The expected convolution product has been identified by measurements of the electric dichroism of DNA restriction fragments with a high time resolution.[32] The dichroism rise curves required two exponentials for a quantitative description according to

$$\Delta A = \Delta A_\infty \left(1 + \frac{\tau_p}{\tau_r - \tau_p} e^{-t/\tau_p} - \frac{\tau_r}{\tau_r - \tau_p} e^{-t/\tau_r} \right) \tag{4.8}$$

where τ_p is the time constant of polarization, which corresponds to the time constant of dipole generation, and τ_r is the time constant of rotational diffusion of the double helices.

The time constants have been measured at different field strengths and at different concentrations of monovalent ions. The reciprocal time constant of polarization $1/\tau_p$ measured at any given field strength E is a linear function of the ion concentration c_{Na} (cf. Fig. 4.11). This result demonstrates that the redistribution of ions induced by electric field pulses does not represent some simple shift of the ions along the double helix, but involves a dissociation reaction according to

$$\text{Na} + \text{DNA} \underset{k^-}{\overset{k^+}{\rightleftharpoons}} \text{Na} \cdot \text{DNA} \tag{4.9}$$

The experimental data show that the rate constant of association k^+ is independent of the electric field strength. In the case of a DNA with 76 base pairs, the association rate constant is $k^+ = 1.1 \times 10^{10}\,\text{M}^{-1}\,\text{s}^{-1}$; a closely corresponding value ($k^+ = 8 \times 10^9\,\text{M}^{-1}\,\text{s}^{-1}$) was found for a DNA with 95 base pairs. These results show that the association of Na^+ ions is a diffusion-controlled process, as should be expected.

The rate constant of dissociation k^- increases with increasing field strength and thus demonstrates that electric field pulses induce a net dissociation of ions from the DNA. This effect is expected, because electric fields are known to induce dissociation of ion complexes. Generation of an electric dipole requires biased dissociation, that is, different degrees of ion association at the ends of the double helix. Actually,

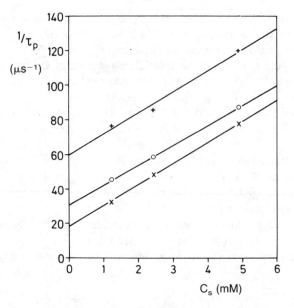

Figure 4.11 Reciprocal polarization time constant $1/\tau_p$ for a DNA restriction fragment with 76 base pairs as a function of the Na^+ concentration at three different electric field strengths ($\times$) 24.6 kV/cm, ($\bigcirc$) 54.1 kV/cm, and (+) 68.6 kV/cm (20°C).

the generation of the dipole moment is the driving force for the biased dissociation, which is demonstrated by the observed linear dependence of $\log(k^-)$ on the square of the electric field strength (E^2 (cf. Fig. 4.12). Extrapolation of the measured dependence of $\log(k^-)$ on E^2 to $E^2 = 0$ provides a value $k^- = 1.3 \times 10^7 \text{ s}^{-1}$ for a double helix with 76 base pairs. This rate constant should be regarded as an average value, which represents dissociation of Na^+ ions from a double helix at Na^+ concentrations of a few millimolar.

In summary, relaxation measurements demonstrate that binding of Na^+ to double-helical DNA is a "simple" reaction with a rate controlled by diffusion.

4.8 Discussion

Early studies of ion binding reactions demonstrated that alkali and alkaline earth ions bind to polynucleotides mainly via electrostatic interactions[33–35]; that is, these ions are attracted by the negative charges of the phosphate residues. Another group of ions is represented by the transition metal ions, which are not only attracted by the highly negative electrostatic potential but also form coordination contacts with base residues. In the present contribution, a small selection of binding reactions has been described for Mg^{2+}; for comparison, some reactions of Ca^{2+} and of Na^+ are included. Binding of these ions to polynucleotides affect their electrostatic potential, and thus a quantitative description of the binding equilibrium requires poly-

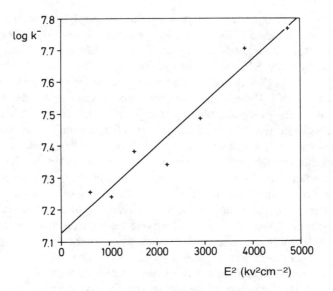

Figure 4.12 Logarithm of the rate constant $\log(k^-)$ for the dissociation of Na⁺ ions from a DNA restriction fragment with 76 base pairs as a function of the square of the electric field strength E^2. The straight line was obtained by linear regression (from Ref. 32).

electrolyte theory. Polyelectrolyte theory has been developed at various levels. A particularly useful form has been given by Manning,[1] which is based on the concept of ion condensation: Ions are "condensed" to the polymer, until the charge density of the polymer is reduced below a critical value. The theory has been developed for the limiting case of an infinitely long line charge. A large number of experimental data obtained for polynucleotides proved to be consistent with the theory.[1]

The condensed ions may be localized at some binding sites or delocalized depending on the nature of the charged group on the polymer and the counterion species. An unequivocal identification of the binding mode into the limit cases of "site binding" and "ion atmosphere binding" proved to be difficult in many cases, where a simple spectroscopic label could not be used for a direct characterization. As shown by many investigations and controversial discussions in the literature, the identification of the binding mode is particularly difficult in the case of the interactions between Mg²⁺ and polynucleotides. As described, a relatively simple approach to the problem is provided by kinetic procedures: When the kinetics of the Mg²⁺-binding reaction can be measured at a sufficient time resolution, the mode of binding can be identified quantitatively.

Another approach to the problem is NMR spectroscopy.[36–44] Although NMR spectroscopy may provide a wealth of different information, the assignment and the interpretation of the data obtained for interactions of simple ions with polynucleotides is apparently not always simple. NMR has been used, for example, to characterize binding of Na⁺ ions to DNA double helices. It has been shown by ²³Na spectroscopy that "the number of associated sodium ions per phosphate charge is

insensitive to large changes in the concentration of NaCl."[43] The competition between different ion species for association to DNA has been described by an ion-
exchange reaction model.[41,42] The mode of Mg^{2+} binding to polynucleotides has
been studied by ^{25}Mg NMR spectroscopy. Rose et al.[36] concluded that the interaction of Mg^{2+} with DNA double helices is dominated by site binding effects. According to Rose et al. "the magnesium ion loses a water molecule from the first
coordination sphere" upon site binding. Berggren et al.[37] studied ^{25}Mg spin relaxation and concluded that "the lifetimes of bound magnesium and calcium may well
be of similar magnitude, although our calculations cannot exclude a significantly
shorter lifetime for calcium." Berggren at al.[37] also concluded that for both ions at
very low amounts of added ion "the interaction cannot be described by simple
electrostatic association, but becomes dominated by some specific interaction."
However, Berggren et al.[37] do not specify the type of interaction. Recently Cowan
et al.[40] also used ^{25}Mg NMR spectroscopy and concluded that "the low value of
the quadrupole coupling constants is indicative of outer sphere coordination by
$Mg(H_2O)_6^{2+}$." Furthermore, Cowan et al.[40] concluded that "G/C-DNA binds
$Mg^{2+}(aq)$ up to 40 to 100-fold more strongly than A/T-DNA." Evidence for a
dependence of Mg^{2+} binding to double-helical DNA on the base pair composition
has not been described in previous investigations. Because of contradictions in the
conclusions drawn from NMR investigations, it seems that some of the criteria used
in these investigations are not as stringent as required. However, owing to the high
potential of NMR spectroscopy, it is expected that further developments and future
applications of NMR techniques will provide more detailed information on the
interactions with polynucleotides. It should be mentioned that the concentrations
used for NMR investigations are usually much larger than those used for the measurements of chemical relaxation described. Thus some differences in conclusions
may be due to differences in the concentrations resulting in a different state of
association.

Although the kinetics of ion binding to polynucleotides has been studied for a
limited number of cases yet, the available data provide a rather clear picture of the
main types of binding modes and of their dynamics. As should be expected according to theory, the "simple" mode of binding into the "ion atmosphere" of polynucleotides is controlled by diffusion. The exact magnitude of a diffusion-controlled
reaction of a positively charged ligand with a polymer characterized by a highly
negative charge density cannot be predicted as accurately as for simple reactants,
which may be described as point charges. The numerical value also depends on the
concentration units used in the evaluation. Nevertheless, it is obvious that rate
constants in the range around 10^{10} M^{-1} s^{-1} indicate a diffusion-controlled reaction.
Rate constants of this order of magnitude have been observed for the binding of Na^+
to double-helical DNA, for example. In this case of a simple alkali metal ion, there
is hardly any objection to the view that "ion atmosphere" binding is the dominant
mode of binding.

A more remarkable observation is the high rate constant of about 10^{10} M^{-1} s^{-1}
for the reaction of Mg^{2+} with several single-stranded polynucleotides, which again
indicates an ion atmosphere type of binding. This high rate constant has been

observed under experimental conditions, where the affinity of Mg^{2+} ions to poly-nucleotides is extremely high, indicated by apparent equilibrium constants of about $10^7 M^{-1}$. Under these conditions it might be expected that there are direct contacts between Mg^{2+} ions and phosphate residues. However, the absence of a slow relaxation effect, which is characteristic of inner-sphere complexes with a direct contact between Mg^{2+} and any ligand, demonstrates the absence of such contacts to phosphate residues of the polynucleotide chain.

An explanation for the dominance of outer-sphere binding of Mg^{2+} to poly-nucleotide chains is provided by the structure of the reactants. In standard structures of polynucleotides, nucleotide residues are stacked upon each other at a distance of 0.34 nm. Furthermore, subsequent residues are rotated with respect to each other by about 36°, leading to a distance between subsequent phosphate residues of 0.7 to 0.8 nm. This distance is too large for simultaneous direct contacts of Mg^{2+} to adjacent phosphates (cf. Fig. 4.13). Alternatively, Mg^{2+} ions may form a direct contact with a single phosphate residue. However, in this case the Mg^{2+}–phosphate complex would be associated with a local positive charge, which appears to be unfavorable. Under these conditions, an arrangement with a minimal electrostatic free energy may be represented by a Mg^{2+} position at the center between two phosphates. It may be as well that the minimum of the electrostatic free energy at the position between two adjacent phosphates is rather shallow. Then Mg^{2+} ions are hardly localized at all and form a highly mobile ion atmosphere around the poly-nucleotide.

According to most available experimental data, Mg^{2+} binding to the standard stacked conformation of straight polynucleotide chains is mainly determined by electrostatic interactions. However, coordination effects involving other types of interactions may contribute, whenever the polynucleotide chain is folded into special structures, which provide specific sites for individual interactions. A partic-

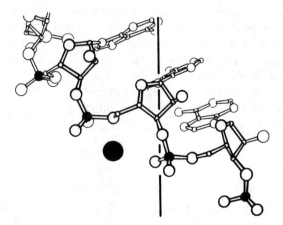

Figure 4.13 Distance between phosphate residues in a ribonucleotide chain with stacked base residues and the size of a Mg^{2+} ion without hydration.

ularly useful example is the anticodon loop of tRNA[Phe], because in this case both the structure in the crystalline state[27-31] and the dynamics of this structure in solution together with the dynamics of Mg^{2+} binding[23,26] have been characterized. The data obtained for the structure and the dynamics are completely consistent and complement each other. The inner-sphere contact site of Mg^{2+} deduced from the dynamics has been identified in the structure as the phosphate of the nucleotide residue 37. The large change of the Wye base fluorescence is explained by a direct contact to Mg^{2+} and by formation of a network of additional contacts, which restricts the conformational flexibility. Moreover, the low rate of inner-sphere complexation appears to be due to a high activation entropy associated with the large number of contacts required for the complex and the restriction of conformational freedom. Finally, the Mg^{2+}-induced transition to one of two conformational states derived from the kinetic data is also explained by the network of Mg^{2+} contacts.

The Mg^{2+} complex formed with the anticodon loop of tRNA[Phe] (yeast) illustrates the conditions for inner-sphere complexation. Apparently a Mg^{2+} contact to a single phosphate is not sufficient for inner-sphere complexation, but additional contacts are required for stabilization. These additional contacts can hardly be provided by straight polynucleotide chains in their standard stacked conformation. The formation of inner-sphere complexes with short oligoriboadenylates indicates that these oligomers are more flexible than the corresponding polymers and are able to display contacts sites in an orientation, which is favorable for coordination. According to the available results, this capacity is dependent both on the base residue and the sugar moiety. The specificity of Mg^{2+} inner-sphere complexation may be important for recognition of specific nucleotide sequences. The particularly long lifetime of Mg^{2+}–inner-sphere complexes may be essential for this purpose.

In summary, Mg^{2+} ions do not only form complexes with polynucleotides based on "simple" electrostatic attraction. Although the electrostatic interactions are dominant in most cases, there are also other types of interactions, which may contribute to complex formation. Due to the special structure of polynucleotides, these additional interactions are required for inner-sphere complexation. The additional contacts can be provided only at specific sites with a particular folded structure of the polynucleotide chain. Formation of these folded structures requires specific nucleotide sequences. Due to the low rate of substitution into the inner hydration sphere of Mg^{2+}, complexes formed at these sites can be identified by investigations of the kinetics. The long lifetime of Mg^{2+}–inner-sphere complexes should be useful for a biological function as "markers," which may be recognized by specific proteins, for example.

References

1. Manning, G. S., *Quart. Rev. Biophys.* **1978**, *11*, 179–246.

2. Record, M. T.; Anderson, C. F.; Lohman, T. M., *Quart. Rev. Biophys.* **1978**, *11* 103–78.

3. Kretsinger, R. H., in *The Neurosciences,* Fourth Study Program, Eds. Schmitt, F. O.; Worden, F. G., MIT Press, 1979, pp. 617–22.

4. Diebler, H.; Eigen, M.; Hammes, G. G., *Z. Naturf.* **1960**, *15b*, 554–60.

5. Diebler, H.; Eigen, M.; Ilgenfritz, G.; Maass, G.; Winkler, R., *Pure Appl. Chem.* **1969**, *20*, 93–115.

6. Marcus, Y., *Ion Solvation*, Wiley & Sons, New York, 1985.

7. Burgess, J., *Metal Ions in Solution*, Wiley & Sons, New York, 1978.

8. Martell, A. E.; Smith, R. M., *Critical Stability Constants*, vols. 1 and 3, Plenum Press, New York, 1974 and 1976.

9. Sillen, L. G.; Martell, A. E., *Stability Constants of Metal-Ion Complexes*, Supplement No. 1, The Chemical Society, Spec. Publ. No. 25, London, 1971.

10. Eigen, M.; DeMaeyer, L., "Relaxation methods," in *Techniques of Organic Chemistry*, Vol. VIII, part II, Weissberger, A., Ed., 1963.

11. Bernasconi, C. F., *Relaxation Kinetics*, Academic Press, New York, 1976.

12. Neely, J.; Connick, R., *J. Amer. Chem. Soc.* **1970**, *92*, 3476–78.

13. Diebler, H.; Secco, F.; Venturini, M., *J. Inorg. Biochem.* **1991**, *42*, 67–77.

14. Hynes, M. J.; Diebler, H., *Biophys. Chem.* **1982**, *16*, 79–88.

15. Peguy, A.; Diebler, H., *J. Phys. Chem.* **1977**, *81*, 1355–58.

16. DeMaeyer, L.; Persoons, A., in *Techniques of Chemistry VI*, Hammes, G. G., Ed., Wiley & Sons, 1974, pp. 211–35.

17. Labhart, H., *Chimia* **1961**, *15*, 20–26.

18. Dourlent, M.; Hogrel, J. F.; Helene, C., *J. Amer. Chem. Soc.* **1974**, *96*, 3398–406.

19. Meyer-Almes, F. J.; Porschke, D., *Biochemistry* **1993**, *32*, 4246–53.

20. Marcandalli, B.; Winzek, C.; Holzwarth, J. F., *Ber. Bunsenges. Phys. Chem.* **1984**, *88*, 368–74.

21. Porschke, D., *Biophys. Chem.* **1976**, *4*, 383–94.

22. Porschke, D., *Nucleic Acids Res.* **1979**, *6*, 883–98.

23. Labuda, D.; Porschke, D., *Biochemistry* **1982**, *21*, 49–53.

24. Urbanke, C.; Maass, G., *Nucleic Acids Res.* **1978**, *5*, 1551–60.

25. Fuller, W.; Hodgson, A., *Nature* **1967**, *215*, 817–21.

26. Bujalowski, W.; Greaser, E.; McLaughlin, L. W.; Porschke, D., *Biochemistry* **1986**, *25*, 6365–71.

27. Jack, A.; Ladner, J. E.; Rhodes, D.; Brown, R. S.; Klug, A., *J. Mol. Biol.* **1977**, *111*, 315–28.

28. Holbrook, S. R.; Sussman, J. L.; Warrant, R. W.; Church, G. M.; Kim, S. H., *Nucleic Acids Res.* **1977**, *4*, 2811–20.

29. Quigley, G. J.; Teeter, M. M.; Rich, A., *Proc. Natl. Acad. Sci. USA* **1978**, *75*, 64–68.

30. Hingerty, B.; Brown, R. S.; Jack, A., *J. Mol. Biol.* **1978**, *124*, 523–34.

31. Teeter, M. M.; Quigley, G. J.; Rich, A., in *Nucleic Acid Metal Ion Interactions*, Spiro, T. G., Ed., 1980, Wiley, New York, pp. 147–77.

32. Porschke, D., *Biophys. Chem.* **1985**, *22*, 237–47.

33. Zimmer, C., *Zeitschrift für Chemie* **1971**, *11*, 441–58.

34. Izatt, R. M.; Christensen, J. J.; Rytting, J. H., *Chem. Rev.* **1971**, *71*, 439–81.

35. Swaminathan, V.; Sundaralingam, M., *CRC Critical Rev. Biochem.* **1979,** *6,* 245–336.

36. Rose, D. M.; Bleam, M. L.; Record, M. T. Jr.; Bryant, R. G., *Proc. Natl. Acad. Sci. USA* **1980,** *77,* 6289–92.

37. Berggren, E.; Nordenskiöld, L.; Braunlin, W. H., *Biopolymers* **1992,** *32,* 1339–50.

38. Strzelecka, T. E.; Rill, R., *Biopolymers* **1990,** *30,* 803–14.

39. Reid, S. S.; Cowan, J. A., *Biochemistry* **1990,** *29,* 6025–32.

40. Cowan, J. A.; Huang, H. W.; Hsu, L. Y., *J. Inorg. Biochem.* **1993,** *52,* 121–29.

41. Braunlin, W. H.; Anderson, C. F.; Record, M. T., *Biopolymers* **1986,** *25,* 205–14.

42. Paulsen, M. D.; Anderson, C. F.; Record, M. T., Jr., *Biopolymers* **1988,** *27,* 1249–65.

43. Bleam, M. L.; Anderson, C. F.; Record, M. T., Jr., *Biochemistry* **1983,** *22,* 5418–25.

44. Black, C. B.; Cowan, J. A., *J. Amer. Chem. Soc.* **1994,** *116,* 1174–78.

CHAPTER 5

Magnesium as the Catalytic Center of RNA Enzymes

Drew Smith

5.1 Introduction

RNA science inevitably becomes magnesium science. From the earliest studies on tRNA structure and stability, to the current interest in RNA catalysis, divalent cation type and concentration ineluctably influence the outcome and interpretation of experiments. Because RNA is a polyanion, its interaction with divalent cations is close and crucial. To describe ribozymes as metalloenzymes, Yarus invoked the image of the Cheshire Cat's smile: The RNA cat fades away, leaving metallic teeth to do the ribozyme's catalytic business.[1] To this metaphor for the catalytic role of magnesium, we can add that of a pursestring for the structural role. Magnesium serves to draw RNA elements together into a tightly ordered structure, capable of holding a catalytic or substrate binding center.

Although catalytic RNAs are the focus of this review, I have included tRNA in the discussions of metal binding. Because we have so much more structural information about tRNA than any other natural RNA, it serves as a Rosetta stone for translating experimental data into RNA structure.

5.2 Stabilization of RNA Structure

There are two somewhat overlapping themes in the structural interactions of Mg^{2+} with RNA: (1) as a specific ligand, contacting several functional groups, and stabilizing a precise conformation; (2) as a counterion, neutralizing electrostatic repulsion. The first theme is distinguished from the second by tight ($\leq 10^{-4}$ M) dissociation constants, some degree of cation selectivity, and, frequently, cooperativity in binding.[2-4]

5.2.1 Transfer RNA

The ability of cations to stabilize RNA structure has been most extensively studied in tRNA by UV absorbance,[5,6] Mg^{2+}-dependent fluorescence,[7] NMR,[8] modified base fluorescence, and equilibrium sedimentation.[9] The secondary structure and most or all of the tertiary structure of tRNA can form with monovalent cation alone at room temperature; at higher temperatures, divalent cation is required to stabilize tertiary structure. About 5 Mg^{2+} are tightly bound, and 20–25 are weakly bound. Under mildly denaturing conditions (e.g., the presence of quaternary alkyl amines) uptake of Mg^{2+} is cooperative; this property probably reflects formation of tertiary structure by binding of the first Mg^{2+}, which then facilitates the formation of other sites. Cooperative interactions between low-affinity Mg^{2+} sites have also been reported.[10] Under native conditions, Mg^{2+} uptake is usually reported to be sequential and independent.

High concentrations of monovalent cation can act to denature tRNA. This effect has been ascribed to competitive inhibition of Na^+ for Mg^{2+} sites,[11] or electrostatic effects.[5] The latter can be understood as an effect on the dielectric constant of the solvent. As the dielectric constant increases, the free energy of charge–charge interactions is proportionately decreased, leading to weakened binding between Mg^{2+} and phosphate.

5.2.2 Catalytic RNAs

5.2.2.1 Tertiary Folding of a Group I Intron Ribozyme Appears Strongly Dependent on Divalent Cations

Physical studies of the interaction of metal ions with catalytic RNAs have been more limited; most studies have centered on the catalytic effects of cations (below). The folding of the cyclized form of the *Tetrahymena* IVS ribozyme in the presence of 1 M Na^+ versus 0.05 M Na^+/0.01 M Mg^{2+} has been compared by chemical modification and optical melting curves.[12] The structure and stability appear to be similar under both sets of conditions, indicating that high monovalent salt can promote the folding of the RNA.

In a different assay, folding of a linear form of the *Tetrahymena* IVS ribozyme appears to be highly dependent on Mg^{2+}.[13] In this assay, O_2 and Fe^{2+}–EDTA combine to generate hydroxyl radicals that cleave nucleic acids at the sugar–phosphate backbone. Cleavage is insensitive to primary sequence or base pairing, but tertiary folding can shield regions of the polynucleotide from attack. This shielding generates a cleavage pattern that is characteristic of the folding of the molecule. Using this assay, it was found that monovalent cation alone does not support proper tertiary folding. Furthermore, addition of 0.3 M Na^+ to 1.5 mM Mg^{2+} strongly reduces the fraction of properly folded ribozyme, and also reduces catalytic activity. The differences in this result with that of Jaeger et al. are probably due to differences in the nature of the assays.[5] The chemical modifications and UV absorbance used by Jaeger et al.[12] are most sensitive to secondary structure (that is, base pairing),[5] while the radical cleavage reaction is most sensitive to tertiary structure. The ionic requirements of the *Tetrahymena* IVS therefore appear to be similar to those of tRNA: monovalent salt permits the formation of secondary structure, but Mg^{2+} is needed for full formation of tertiary structure.

The Fe^{2+}–EDTA assay also indicates that ribozyme folding is cooperatively dependent on Mg^{2+} concentration. Analysis of the dependence indicates that three or more Mg^{2+} are bound. Catalytic activity, as measured by the parameter k_{cat}/K_M, correlates with folding. Oligonucleotide substrate binding, rather than chemistry, is the rate-limiting step for k_{cat}/K_M,[14] and so the correlation between k_{cat}/K_M and folding presumably reflects the requirement that the ribozyme be fully folded for substrate binding. Ca^{2+} and Sr^{2+} are also effective in promoting folding but do not support catalysis. Mg^{2+} appears to function as a pursestring for this ribozyme, drawing together and stabilizing its structural elements. This type of binding is analogous to the Mg^{2+} site in the core of tRNA, which bridges the D- and TΨC loops, stabilizing their interaction.

5.2.2.2 Other Ribozymes Can Fold in the Absence of Divalent Cations

The binding of eubacterial RNase P RNA to tRNA can be assayed by a cross-linking reaction.[15] An aryl azide is attached to the 5′-phosphate of tRNA, which is the site of cleavage by the ribozyme. Binding of ribozyme to tRNA requires that both be folded correctly. High concentrations of monovalent cation (e.g., 1 M) are optimal for enzymatic activity. At these high monovalent concentrations, the RNase P RNAs from three different eubacteria are able to cross-link tRNAs[Phe] from yeast or *E. coli* at 10–30% yield when the reactants are at approximately K_M concentrations. The sites of cross-linking are identical in the presence or absence of Mg^{2+}. The extent of cross-linking in the absence of Mg^{2+} is 9 to 93% of the extent in the presence of Mg^{2+}, depending on the tRNA–RNase P RNA pair.[16]

A kinetic analysis of the cleavage reaction also demonstrates pre-tRNA binding in the absence of Mg^{2+}. Pre-tRNA and RNase P RNA were incubated together in the absence of Mg^{2+}. Upon addition of Mg^{2+}, the time course of product formation shows an initial burst, followed by a slower, steady-state phase of cleavage. The burst of product formation is interpreted as rapid cleavage of substrate already bound, in the absence of Mg^{2+}, to the ribozyme.[16,17] A similar experiment has been performed for the hairpin ribozyme, with similar results.[18]

There is evidence for Mg^{2+}-independent folding of the hammerhead ribozyme. The NMR spectrum of one variant of the hammerhead ribozyme shows no significant differences in the presence and absence of Mg^{2+}.[19] However, this experiment was performed with a ribozyme containing a 2′-deoxyribose substitution at the cleavage site, which renders the substrate uncleavable. This substitution could also disrupt Mg^{2+} binding. Another noncleavable analog of a hammerhead ribozyme appears to take up a single Mg^{2+}, as determined by the Mg^{2+}-dependent change in the ellipticity of its CD spectrum.[20] The ellipticity indicates a conformational change. This analog, which incorporates a 2'-O-methyl substitution at the cleavage site and confers resistance to cleavage, may not bind Mg^{2+} in the same mode as a catalytically active ribozyme, however.

RNase P RNA, the hairpin, and possibly the hammerhead ribozymes appear to fold correctly in the absence of divalent cation, if other cations are supplied. These monovalent cations are able to provide sufficient counterion shielding of charge repulsion between phosphates to permit folding. Because these RNA molecules form tertiary structures in the absence of Mg^{2+}, it follows that their Mg^{2+} sites need not consist of complex tertiary structures. Instead, their Mg^{2+} sites may consist of small, autonomous structures, such as bulges and loops. These types of sites are especially likely for the small ribozymes. The anticodon loop of tRNA provides an example of a hairpin loop binding site for Mg^{2+}. Mg^{2+} is not necessary for formation of the anticodon loop, but fine tunes and stabilizes its conformation. Like tRNA, the forces of base stacking and hydrogen bonding (with electrostatic repulsion neutralized by monovalent cation) may be sufficient to fold these RNAs into a native or near-native structure in the absence of Mg^{2+}.

5.3 Metal Binding Sites on RNA

In the case of tRNA, X-ray crystallography has been used to examine metal-binding ligands on RNA.[21-23] From work with simple complexes[24] it is expected that alkaline earth metals, such as Ca^{2+} and Mg^{2+}, would be able to coordinate directly only to functional groups that carry a negative charge, that is, phosphate oxygens. Transition elements (e.g., Mn, Ni, Zn, Os, Pt) can also coordinate directly to phosphate oxygens, and, in addition, to the weakly basic ring nitrogens of the nucleotide bases. Both types of metals can form outer-sphere complexes by hydrogen bonds to nucleotides through water coordinated to the metal.

Crystallographic analysis of Mg^{2+} binding is difficult. Because Mg^{2+} is a light element, it does not diffract X rays strongly, and is difficult to distinguish from Na^+ or water molecules. Assignment of electron-density peaks to Mg^{2+} therefore relies somewhat on indirect reasoning, such as occupancy of a site that can also be occupied by a heavy metal, or internuclear distances that are characteristic of coordination bonds ($\sim$2 Å) rather than hydrogen bonds ($\sim$3 Å).

Despite these caveats, the number of Mg^{2+} found by crystallography correlates well with the number found by solution techniques, 4—6 per tRNA. It is possible that this correlation is somewhat misleading; only the sites having the highest occupancy and the least amount of movement within the site are likely to be detected. All of the Mg^{2+} detected are found in loop or turn regions, where multiple phosphate ligands are made available by the folding back of the RNA, and they serve to stabilize tertiary structure. Only phosphate oxygens are found to coordinate directly to Mg^{2+} (distance < 2.2 Å), as expected. Hydrogen-bonding partners of Mg^{2+}-coordinated water include phosphate oxygens, ribose 2'-OH,[21] guanosine N7 and O6, uridine O4, and cytosine N3,[22] adenosine N2, uridine O2, and ribose O5'.[23]

5.3.1 Equivalence of Mg^{2+} Binding to Binding of Other Metals

Given its biological importance, it is unfortunate that Mg^{2+} is not very amenable to biophysical analysis. Mg^{2+} does not fluoresce, or give an ESR or NMR signal, or, at neutral pH, act as a strong general base to promote phosphodiester hydrolysis; other metals that have these properties are frequently used in its place. The hope is that these other metals bind equivalently, and can be used to deduce something about the location and properties of Mg^{2+} binding sites. The crystallographic study of Jack et al. compares the binding of several different metals, including heavy metals, lanthanides, transition metals and earths, and Mg^{2+}.[22] Examination of their binding sites provides a good test of their equivalence to Mg^{2+}, and indicates the caveats and limitations that must be applied to these studies.

Overlap between strong Mg^{2+} binding sites and binding sites for other metals is not very frequent. Two of five lanthanides that bind tRNA^{Phe} overlap Mg^{2+} sites. The main Co^{2+} and Mn^{2+} sites are close enough to Mg^{2+} sites to displace Mg^{2+}, but their binding is different (Fig. 5.1). Co^{2+} and Mn^{2+} coordinate directly to N7 groups of guanosines, and form hydrogen bonds through water to phosphates with which Mg^{2+} coordinates directly. In result, these divalent cations occupy the same binding pocket as Mg^{2+}, but are displaced by a few angstroms, and interact with a different set of ligands. These subtle differences may not strongly affect RNA folding and structure, but can have a marked effect on the catalytic function, as will be discussed.

5.3.2 Metal-Binding Sites in Ribozymes—Manganese Rescue

The technique of "Mn rescue" has been applied to the *Tetrahymena* IVS ribozyme to identify sites of Mg^{2+} coordination to phosphates. In these experiments, sulfur is substituted for one of the phosphate oxygens on the RNA backbone, and the effect on catalytic activity is determined. Sulfur is a very poor coordination ligand to Mg^{2+}, and so it is expected that some of the observed catalytic effects are due to the loss or weakening of Mg^{2+} binding sites. These sites are identified by the ability of Mn^{2+} to restore catalytic activity. Mn^{2+} has a smaller relative preference for phosphates over phosphorothioates (that is, for oxygen over sulfur),[25,26] and so can bind to some fraction of the substituted sites, and function well enough to restore activity. This assay cannot distinguish between Mg^{2+} ions whose primary function is to promote folding, from those whose primary function is to promote catalysis.

At least 28 Mg^{2+}-binding phosphates are identified in the *Tetrahymena* IVS ribozyme by this method.[27] Most of these are clustered in the conserved catalytic core of the ribozyme, and several are in a position to interact with the substrate. When mapped onto the Michel–Westhof three-dimensional model,[28] many of the

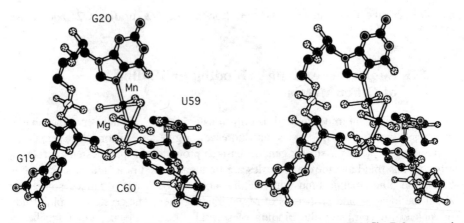

Figure 5.1 Comparison of Mg^{2+} and Mn^{2+} binding in the yeast tRNA^{Phe} D-loop metal binding pocket. Mg^{2+} forms a coordination bond directly to the *pro*-R_p oxygen of G19, and forms hydrogen bonds through coordinated water to G20-N7, G20-O6, U59-O4, and C60-N3. Mn^{2+} coordinates directly to G20-N7, and forms hydrogen bonds through water to G20-O6, the *pro*-S_p oxygen of G20, the *pro*-R_p oxygen of G19, U59-O4 and C60-N3. By binding to sites on both the D loop (nucleotides 19 and 20) and the TΨC loop (nucleotides 59 and 60), the metal ions draw these structures together, stabilizing the tertiary folding of the tRNA. The center–center distance between the cations is ~2 Å. Lead also binds in this pocket, forming coordination bonds to U59-O4 and C60-N3. Hydrogen bonds through water are formed between lead and the G19 *pro*-S_p oxygen, C60-N4, C60-O2, and U59-N3. Mg^{2+}, Mn^{2+}, and Pb^{2+} all therefore occupy the same binding pocket, and have many ligands in common in their respective binding schemes. However, only lead is catalytically active, promoting cleavage of the backbone between residues 17 and 18. The coordinates were obtained from Refs. 22 and 31.

Mg^{2+} appear to be positioned to stabilize the charge repulsion due to the close approach of backbone phosphates, and to stabilize sharp bends in the backbone.

5.3.3 Metal-Catalyzed Hydrolysis of the Phosphate Backbone

The water complexes of divalent cations act as general base catalysts promoting hydrolysis of the RNA phosphodiester backbone. By using cation–water complexes with a low pK_a or raising the pH sufficiently, hydrolysis of an RNA will occur in the vicinity of the cation, helping to localize its binding site. The prototype experiment is the Pb^{2+}-catalyzed cleavage of tRNA^{Phe}.[29–31] Lead binds at three sites; in the D loop it coordinates to phosphate 19, and forms a hydrogen bond through a coordinated water to the 2′OH of nucleotide 17. The reaction results in the cleavage of the phosphate backbone between nucleotides 17 and 18. The cleavage reaction therefore correctly identifies the vicinity of metal binding. Subsequent studies have shown that lead cleavage is a sensitive probe of tertiary structure,[32] as it is very dependent on the proper positioning of its ligands for binding and catalytic activity.

Group I intron ribozymes display two major sites of lead cleavage, both in the phylogenetically conserved core, positions A263 and U305 in the *Tetrahymena* IVS.[33] These sites are near the substrate binding region in the three-dimensional model,[28] and their cleavage is inhibited by 2'dGTP, a competitive inhibitor of the ribozyme reaction. In addition, these sites are identified as Mg-binding sites by the "manganese rescue" experiment.[27]

Mg^{2+} at pH 9.5, as well as other divalent cations, was used to induce cleavage of RNase P RNA.[34] Of five cleavage sites identified in the *E. coli* enzyme, the most interesting is that at position 120, which is phylogenetically conserved.[35] The analogous site in the RNA secondary structure is also cleaved in *B. subtilis* RNase P RNA, indicating that it is a site of general importance.

5.3.4 Specificity of Binding

The work on cation requirements is summarized in Table 5.1. Monovalent cations have also been included under the category of structural ions. Some general trends can be deduced by inspection of these data:

1. Mg^{2+} is the only cation that can support catalysis for all ribozymes. Mn^{2+} and Ca^{2+} each work for four of the five. However, rates tend to be much lower in Ca^{2+}-promoted reactions than in Mg^{2+}- or Mn^{2+}-promoted reactions.
2. The structural sites are usually less discriminating in the type of cation required, and can sometimes accommodate monovalent as well as divalent cations. The reduced specificity in cation requirement at these sites suggests that a primary requirement is charge neutralization between adjacent phosphates. The greater degree of specificity in the catalytic sites could be structural (that is, only certain ions can bind) or chemical (inactive ions bind, but do not promote catalysis). Lead binding to tRNA provides an example of chemical specificity—three lead ions bind, but only one promotes cleavage, presumably because of ineffective positioning of its ligands.
3. Comparison with Table 5.2 shows that there is no simple correlation between the ability to support catalysis and the primary structural and chemical properties of the cations. The range in sizes of the generally active cations (Mg^{2+}–Ca^{2+}, Table 5.2) encompasses much of that of the generally inactive ones (Ba^{2+}– Zn^{2+}, Table 5.2). It is therefore likely that some of the inactive ions are able to bind in the same sites as the active ones, but are unable to promote hydrolysis.
4. Some cations that can support catalysis are not able to support proper folding. Zn^{2+} and Cd^{2+} are able to promote catalytic activity for the hammerhead ribozyme only in the presence of spermine, suggesting that the RNA folds incorrectly with these divalent cations alone.[36] Mn^{2+} and Co^{2+} have been shown to inhibit substrate binding in the hairpin ribozyme reaction.[37] These findings suggest that transition elements are able to denature the RNA. Unlike the alkaline earths (e.g., Mg^{2+} and Ca^{2+}), these elements are able to coordinate with weakly basic sites, such as purine N7, which could lead to inappropriate folding of the RNA.
5. There is no apparent relationship between ribozyme size (and hence, structural complexity) and cation selectivity. The small, simple delta agent and hairpin ribozymes are as selective as the *Tetrahymena* IVS and RNase P ribozymes,

Table 5.1 IONIC REQUIREMENTS OF RIBOZYMES

Ribozyme	Catalytic Ions		Structural Ions	
	Active[a]	Inactive	Enhancement	No Enhancement
Tetrahymena IVS	Mg^{2+}, Mn^{2+}	Ca^{2+}, Sr^{2+}, Ba^{2+}, Zn^{2+}, Co^{2+}, Cu^{2+}	Mn^{2+}, Ca^{2+}, Sr^{2+}, Ba^{2+}	Zn^{2+}, Co^{2+}, Cu^{2+}, Na^+, K^+
RNase P RNA	Mg^{2+}, Mn^{2+}, Ca^{2+}	Sr^{2+}, Ba^{2+}, Zn^{2+}, Co^{2+}, Cu^{2+}, Fe^{2+}, Ni^{2+}, Sm^{2+}	Mn^{2+}, Ca^{2+}, Sr^{2+}, Ba^{2+}, NH_4^+, Na^+, K^+, Rb^+, Cs^+	Zn^{2+}, Co^{2+}, Cu^{2+}, Fe^{2+}, Ni^{2+}, Sm^{2+}, Li^+, $N(CH_3)_4^+$
Hammerhead	Mn^{2+}, Mg^{2+}, Ca^{2+}, Co^{2+}, Sr^{2+}, Ba^{2+}, Zn^{2+}, Cd^{2+}	Pb^{2+}		
Hairpin	Mg^{2+}, Ca^{2+}, Sr^{2+}	Mn^{2+}, Co^{2+}, Cd^{2+}, Ni^{2+}, Ba^{2+}		Mn^{2+}, Co^{2+}
Delta	Mg^{2+}, Mn^{2+}, Ca^{2+}		NH_4^+, Na^+, K^+	

[a] Listed in approximate order of activity. References: *Tetrahymena* IVS,[13,45] RNase P RNA,[43,52,71] Hammerhead,[36] Hairpin,[37] Delta.[72,73]

Table 5.2 PROPERTIES OF DIVALENT CATIONS

Cation	Ionic radius (Å)[a]	H_2O–M^{2+} distance (Å)[b]	Hydration enthalpies (kcal/mol)	Rate of H_2O Exchange (s^{-1})	pK_a	Stability constant, $[M^{2+}$:AMP]/ $[M^{2+}][AMP]$[c]
Mg	0.65	2.1	−459	6×10^5	11.4	93
Mn	0.80	2.2	−441	2×10^7	10.0	250
Ca	0.99	2.4	−380	3×10^8	12.6	71
Sr	1.13	(2.5)	−346		13.1	62
Ba	1.35	(2.7)	−312	1×10^9	13.3	54
Co	0.78	2.1	−491	2×10^6	8.0	440
Cu	0.69	2.0	−502	3×10^9	8.1	1500
Ni	0.62	2.0		5×10^4	10	690
Zn	0.74	2.1	−489	5×10^7	9.5	520

[a] Crystal radii.
[b] By X-ray diffraction in water, except in parentheses, which are approximated by the sum of van der Waals radii for oxygen and the metal.
[c] By pH titration. References: 24, 74, 75.

which are some ten times as large. This implies that local structures (e.g., loops and bulges) are able to form specific metal-binding sites; numerous long-range interactions and higher-order structures are not required. The specificity of the small ribozymes is all the more striking when one considers that the reaction they catalyze, hydrolysis to leave (2′, 3′) phosphate, is promoted non-specifically by divalent cations. Only the hammerhead ribozyme appears to provide a general metal-binding site to promote hydrolysis. As noted, this selectivity can be both structural (inactive ions cannot bind) or catalytic (inactive ions can bind, but are chemically inert).

5.4 Catalysis

5.4.1 Types of Reaction Catalyzed

Ribozymes, in general, catalyze[a] the hydrolysis, condensation, or transfer of phosphate esters. These reactions are the best studied, especially with respect to participation by divalent cations, and will be extensively discussed in the following. Other reactions merit mention. There is good evidence that 23S rRNA catalyzes the peptidyl transfer reaction.[38] Deproteinized ribosomes retain the ability to transfer an N-acetylated amino acid from a tRNA to the aminoacyl–tRNA analog puromycin.

---

[a] Many ribozymes, strictly speaking, are not catalysts, as they are transformed over the course of their reactions, for instance, in self-cleavage. Most of these, however, have been shown to function as true catalysts when engineered to act on an exogenous substrate. Restriction of the term "catalysis" is therefore not especially useful, and I will use it to refer to any RNAs which accelerate a reaction rate.

In contrast to the phosphate ester reactions of ribozymes, the attack of a nucleophilic amine on an acyl carbon is catalyzed in this reaction. Mg^{2+} is required, but its participation in the reaction mechanism is currently only a plausible assumption. In a chemically similar reaction, a modified *Tetrahymena* IVS ribozyme weakly catalyzes the hydrolysis of an aminoacyl–RNA ester[39]; this reaction differs from the peptidyl transfer reaction in that water, rather than an amine, is the attacking nucleophile. Mg^{2+} is also required for this reaction; its participation in the reaction mechanism is inferred from that of other reactions catalyzed by this ribozyme,[40] as will be discussed.

The rationale for the preponderance of phosphate ester catalysts among natural ribozymes is not clear. RNAs clearly are able to bind a variety of metal ions with high specificity and affinity; this property should enable them to perform a number of types of reactions.[1] In vitro "evolution" experiments have shown that RNAs are also capable of binding a variety of large and small molecules, with K_d values in the range of 10^{-10}–10^{-6} M. These two properties should combine to equip RNA with the tools of a versatile and efficient catalyst.

5.4.2 Participation of Mg^{2+} in Catalysis

The first proposals for ribozyme catalytic mechanisms invoked the participation of Mg^{2+}.[41,42] This participation was generally accepted on the grounds of a divalent cation requirement, by analogy to the mechanisms of protein nucleases and phosphatases, and, perhaps most compellingly, by the necessity of a general base catalyst. Unmodified nucleotides have no functional groups that, at physiological pH, can plausibly act in this latter role—however, it is possible that folding of an RNA can produce perturbed pK_a values, as is the case with protein enzymes.

Much of the work that defined cation requirements and optima was done in the absence of a kinetic framework for the reactions. In consequence, the reaction step affected (e.g., substrate binding or chemistry) by changes in type or concentration of metal could not be identified. Substrate binding may occur in a catalytically inactive ribozyme; conversely, a cation that could support the reaction chemistry may not promote optimal folding of the substrate binding site. However, these studies, as well as later ones, showed that divalent cations that were not themselves able to support catalysis could supplement the requirement for Mg^{2+}.[13,43–45] These findings led to the notion of two classes of divalent cation binding sites: (1) structural sites, where a divalent cation is needed for proper folding of the RNA; and (2) catalytic sites, which bind divalent cations that participate directly in the reaction mechanism.

The weakness of this type of evidence is that divalent cations could be involved only indirectly in catalysis. That is, Mg^{2+} could be required for proper folding of the RNA into a catalytically active conformation without actually participating in the reaction mechanism. A plausible basis for this skepticism is provided by reports of metal-ion-independent hydrolysis.[36,37] This possibility has been addressed by experiments indicating the proper folding of ribozymes in the absence of Mg^{2+}.[16,17,38,46] Since Mg^{2+} is required for catalysis, evidence that ribozymes can fold and properly bind substrate in the absence of Mg^{2+} implies that the Mg^{2+} requirement for cleavage is primarily chemical, and not structural.

More direct evidence of Mg^{2+} participation in a ribozyme reaction mechanism is provided by a "manganese rescue" experiment.[40] The $3'$ oxygen (the leaving group) at the substrate cleavage site in the *Tetrahymena* IVS reaction was replaced by sulfur. In the presence of Mg^{2+}, the chemical step of the cleavage reaction is reduced 1000-fold compared to $3'O$. When Mn^{2+} replaces or is added to Mg^{2+}, the chemical step of cleavage of $3'S$ is reduced less than tenfold compared to $3'O$. The correlation of catalytic rates with the coordination preferences of Mg^{2+} and Mn^{2+} for sulfur and oxygen strongly implies the coordination of Mg^{2+} to the $3'O$ at the cleavage site.

5.4.3 Probable Mechanisms; Phosphoryl Transfer Chemistry

Before beginning a discussion of the role of metal ions in catalysis, it will be useful to review the features of biological phosphate ester transfer, as these define the opportunities for metal ion participation. In principle, phosphoryl transfer can occur by either a *dissociative* (S_N1-type), or *associative* (S_N2-type) mechanism. In the former, elimination of the leaving group is the first step, producing a free, unstable metaphosphate intermediate. The second step is rapid addition of the nucleophile [Fig. 5.2(a)]. It is generally believed that enzymes employ the associative mechanism, in which addition of the nucleophile to form a pentacovalent intermediate is first, followed by elimination of the leaving group [Fig. 5.2(b)].

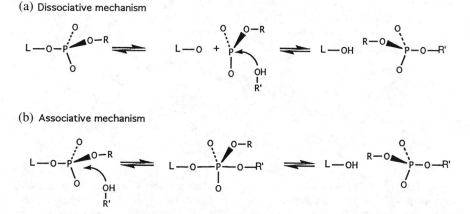

(a) Dissociative mechanism

(b) Associative mechanism

Figure 5.2(a) Dissociative mechanism of phosphoryl transfer. The leaving group (L) dissociates from phosphate to generate a trigonal metaphosphate intermediate. This intermediate is highly susceptible to nucleophilic attack, which regenerates the tetrahedral phosphate center. **(b)** Associative, in-line mechanism of phosphoryl transfer. The reaction is initiated by attack of the nucleophile at the position apical to the leaving group (L). A pentacovalent, trigonal bipyramidal intermediate is formed. This intermediate is resolved to tetrahedral geometry, with inversion of configuration, upon the departure of the leaving group.

The associative mechanism can be subdivided further into *adjacent* and *in-line* mechanisms. In the former, the nucleophile attacks on the same side of the phosphorus as the leaving group, in the latter on the opposite side [Fig. 5.2(b)]. It is generally accepted that biological phosphate transfer occurs by the in-line mechanism. A consequence of the in-line mechanism is that there is inversion of configuration about the phosphorus center. The geometry of the intermediate is a trigonal bipyramid, in which the attacking and leaving groups occupy the apexes, and the phosphate oxygens occupy the equatorial positions. The starting point for all discussions of ribozyme-catalyzed phosphoryl transfer, then, has been the hypothesis that the reaction proceeds through an associative, in-line mechanism. This hypothesis is supported by evidence for inversion of configuration in both ligation and cleavage reactions of the *Tetrahymena* IVS ribozyme,[47,48] and the hammerhead ribozyme reaction.[49]

Ribozyme reactions can also be classified according to leaving group. The 3' hydroxyl constitutes the leaving group in reactions catalyzed by group I and II introns and RNase P RNA [Fig. 5.3(a)]. The 5' hydroxyl leaves in reactions catalyzed by the hammerhead, hairpin, and delta agent ribozymes [Fig. 5.3(b)].

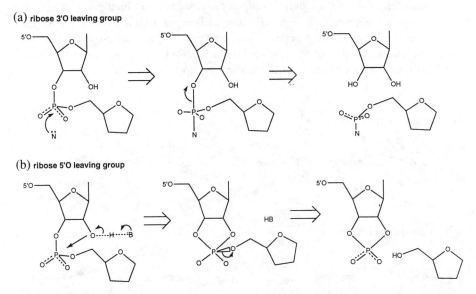

Figure 5.3(a) Mechanism for 3' leaving group. The reaction is initiated by attack of a nucleophile (N) on the phosphate, resulting in formation of the pentacovalent trigonal bipyramidal intermediate. Electron flow from the 3'O–P bond to the 3'O results in the buildup of negative charge on the 3'O leaving group. Since this 3'O anion is itself highly nucleophilic, neutralization of its charge is critical to prevent reversal of the reaction. (b) Mechanism for 5' leaving group. The reaction is initiated by abstraction of a proton from the ribose 2'OH by a general base (B). The 2'O is made highly nucleophilic by electron flow to it from the O–H bond. The 2'O⁻ attacks phosphorus, generating a pentacovalent trigonal bipyramidal intermediate. The intermediate decomposes by expulsion of the 5' leaving group, resulting in a 2',3' cyclic phosphate. This cyclic phosphate spontaneously decomposes slowly by hydrolysis to a mixture of 2' and 3' monophosphates.

5.4.4 Mechanism of Cleavage—5' Hydroxyl Leaving Group

This reaction, catalyzed by the hammerhead, hairpin, and delta agent ribozymes, probably has much in common with two well-characterized reactions: lead cleavage of tRNA and RNase A cleavage. All hydrolyze RNA to leave 5'-OH and cyclic 2',3' phosphate. Crystallographic studies of the Pb^{2+}-catalyzed cleavage reaction[30,31] are particularly relevant to the ribozyme reactions. In these hydrolysis reactions, the first step is activation of the 2'-OH by deprotonation to form the alkoxide, which is a strong nucleophile. The general base that accomplishes this deprotonation is a His residue in RNase A, or the Pb–OH-complex in the lead reaction. In the second step, the 2'-O-nucleophile attacks phosphorus, forming a pentacovalent trigonal by-pyramidal intermediate. The RNase A reaction involves stabilization of the intermediate by the electrostatic interaction of a lysine ϵ-NH_3^+ with the phosphate oxygen. Although coordination by Pb^{2+} could accomplish the same stabilization, it appears to be too far away from the phosphate oxygen to perform this role. In the third step, the 5'-O-leaving group departs—this leaving is stabilized in the RNase reaction by a second His acting as a general acid to donate a proton. In the lead reaction, the distance between Pb^{2+} and the 5'-O leaving group (6 Å) is appropriate for a lead-bound water to act as the proton donor. Assuming that this donation occurs, the ionization state of the Pb-aquo complex is regenerated:

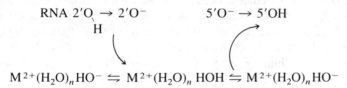

The product of both the RNase A and lead reactions is cyclic 2',3' phosphate, which will slowly hydrolyze spontaneously to form a mixture or 2'- and 3'-monophosphates.

Several opportunities for the contribution of Mg^{2+} to catalysis can be surmised from this framework. First, a Mg-aquo complex can act as a general base in the form $Mg(H_2O)_nOH-$, to accept a proton from a hydroxyl or water. The effectiveness of Mg^{2+} and other metals in this role is a function of pK_a (Table 5.2). A metal–water complex with a low pK_a is ionized to form a hydroxide to a greater extent than one with a higher pK_a. Second, Mg^{2+} coordination to a phosphate oxygen may increase the polarization of the P–O bond. This coordination will increase electron withdrawal from phosphorus, thereby increasing its susceptibility to nucleophilic attack. Third, Mg^{2+} coordination can enhance the stereochemistry of the reaction by stabilizing the trigonal bipyramidal transition state through which the phosphorus passes as it moves from the leaving group to the nucleophile. Further stereochemical enhancement of reactivity can be obtained by the positioning of functional groups, such as hydroxyls or water, through hydrogen bonds to Mg^{2+}-coordinated water. Fourth, Mg^{2+} can act as a Lewis acid, neutralizing negative charge in the transition state or in reaction intermediates. This function should be especially important in stabilizing the leaving group, which is an oxyanion. Fifth, $Mg(H_2O)_n$ can act as a general acid, donating a proton to the leaving group.

5.4.5 Metal Ions in Catalysis—5′ Leaving Group

The hammerhead ribozyme has been a favorite subject for analysis of metal participation in catalysis. Its small size facilitates the use of chemically altered oligonucleotides in the substrate or catalytic moieties. The effect of these alterations on catalysis is typically measured by analysis of the kinetics of the cleavage reaction. This type of assay is not only convenient, but, by definition, reports the behavior of catalytically active complexes—a condition that is not necessarily true in purely structural assays. The difficulty of kinetic assays is that it is not always clear what they measure. Cleavage rates may be limited by substrate binding or product release, or chemistry may be rate limiting. The identity of the rate-limiting step may change upon chemical alteration of substrate or enzyme, or upon a change in reaction conditions. Although it is sometimes assumed that the Michaelian parameter k_{cat} represents the chemical step of cleavage, this is not always the case. For the bipartite hammerhead reaction, in which two RNAs are annealed to form the ribozyme, the chemical step of the reaction is comparable to k_{cat}.[50] For intramolecular cleavage reactions, such as those of the hairpin and delta agent ribozymes, there are no substrate binding or product release steps; the observed rates in these reactions must represent the rates of chemistry, or of a conformational rearrangement prior to cleavage. Therefore, for the reactions in which the leaving group is the ribose 5′OH (the hammerhead, delta agent, and hairpin ribozyme reactions), large effects on k_{cat} (for bimolecular reactions) or k_{obs} (for intramolecular reactions) are usually attributable to effects on the chemical step of the reaction. For reactions in which the ribose 3′OH is the leaving group, this is not usually the case. Product release is rate limiting under steady-state conditions for both RNase P RNA[16,17,46,51,52] and the *Tetrahymena* IVS ribozyme.[14]

5.4.6 Magnesium-Bound Water as a General Base

From the model of Pb-catalyzed cleavage of tRNA, it is expected that hammerhead, hairpin ribozyme, or delta agent cleavage reactions are initiated by abstraction of a proton from the substrate 2′OH by a M^{2+}–hydroxide complex. Since the active metals in these reactions have pK_a values > 9, they exist predominantly as inactive protonated complexes at physiological pH. Cleavage rates should therefore increase proportionately with hydroxide concentration up to the pK_a of the complex, as the water bound to the metal is converted to hydroxide. This first-order dependence on hydroxide concentration has been observed in the hammerhead reaction.[53] As this observation was made over a pH range in which there are no titratable groups, the effect is most plausibly assigned to the deprotonation of the catalytic metal–water complex.

The model in which a M^{2+}–water complex acts as a general base also predicts a relationship between catalytic rates that reflects the relationship between pK_a values of different catalytic metal ions. For the hammerhead ribozyme, the Mn^{2+} ($pK_a = 10$) reaction is about sixfold faster than the Mg^{2+} ($pK_a = 11.4$) reaction, which in turn is 30-fold faster than the Ca^{2+} ($pK_a = 12.6$) reaction. The correlation of the order of reactivites with pK_a, and the rough proportionality of rate and pK_a, are further evidence that a metal–water complex acts as a general base in the reaction.

5.4.7 Coordination to the Substrate Phosphate

In the tRNA Pb^{2+} cleavage reaction, the lead ion does not coordinate to the cleaved phosphate. Instead, there is a long contact to phosphate 19, one nucleotide displaced from the cleavage site. Coordination to the cleaved phosphate therefore plays no role in promoting the reaction. In contrast, a metal ion does appear to coordinate to the scissile phosphate in the hammerhead reaction. Substitution of the hammerhead cleavage site pro-R phosphate oxygen[b] with sulfur increases the Mg^{2+} requirement of the reaction by at least two orders of magnitude.[36] Even at saturating Mg^{2+} concentration, the cleavage rate is reduced several-fold.[20,36] This rate effect is not intrinsic to the chemistry: The thio effect in an analogous nonenzymatic base-catalyzed hydrolysis is very small.[54] Most of the rate reduction can be recovered by addition of Mn^{2+},[36] which, unlike Mg^{2+} can coordinate to sulfur. These data indicate direct coordination to the pro-R phosphate oxygen. By stabilizing negative charge on the oxygen, this coordination could facilitate electron withdrawal from the phosphorus, increasing its susceptibility to nucleophilic attack.

This coordination does not appear to be a universal feature of ribozymes that catalyze reactions in which the 5′ oxygen is the leaving group. Substitution of the pro-R phosphate oxygen by sulfur has little effect on the rate of cleavage by the hairpin ribozyme[18]; substitution at the pro-S phosphate oxygen has not been checked. Given that these reactions involve an intramolecular attack of a potent alkoxide nucleophile on phosphorus, little more may be required than activation of the 2′ hydroxyl by a metal–hydroxide complex. It is probably fair to think of this class of ribozymes as small metal-binding pockets, which have only enough structure to bind a metal ion at physiological concentrations, and direct it toward deprotonating a single 2′OH.

5.4.8 Metal Ions in Catalysis—3′ Leaving Group

In contrast, the reactions promoted by group I intron ribozymes and RNase P RNA are somewhat more complicated, as they require intermolecular nucleophilic attack. It is not surprising then, that the role of metal ions is more complex in these reactions, and that the actions of several metal ions appear to be required for catalytic activity.

---

[b] If one imagines the two non-bridging phosphate oxygens coming out of the plane of the phosphorus toward the viewer, with the RNA chain running 5′ to 3′, top to bottom, the pro-S phosphate oxygen is on the right, and pro-R is on the left. The pro-R oxygen is closest to the ribose 2′OH.

5.4.9 Magnesium-Bound Water as a General Base— Activation of the Nucleophile

Haydock and Allen[41] and Guerrier-Takada et al.[44] proposed a general one-metal mechanism for ribozymes in which $Mg(H_2O)_nOH^-$ acts as a general base[4] accepting a proton to activate the nucleophilic attack of water or guanosine 3'-OH on the substrate

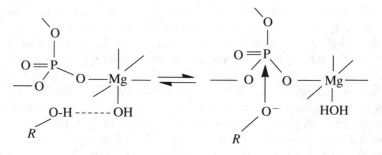

R = H, guanosine

phosphorus. The pK_a of the Mg-aquo complex (11.4) is below that of water (15.7) and ribose (12.5),[55] and so can function as a proton acceptor. However, this pK_a is well above physiological pH, and so only a small fraction ($\sim 10^{-4}$) of the Mg-aquo complexes are expected to be ionized to form active complexes (unless the pK_a is perturbed). This fraction should increase linearly with pH, until the pK_a of the complex is approached. A prediction of this model is that the rate of the chemical step of the reaction should increase logarithmically with pH (that is, linearly with hydroxide concentration), if activation of the nucleophile is rate limiting.

The pH dependence of cleavage rate has been studied for RNase P RNA and the *Tetrahymena* IVS ribozyme.[52,56] In both cases a logarithmic dependence of rate on pH is observed at pH values below 7.5; above this range there is a plateau. In the case of the *Tetrahymena* ribozyme reaction, this plateau probably represents the pH at which substrate binding becomes rate limiting under the experimental conditions used. Although the RNase P RNA reactions were done under conditions where chemistry is expected to be rate limiting, another trivial explanation of the plateau is that it represents denaturation of the ribozyme at high pH. This explanation seems unlikely, as the plateaus begin at pH 8–8.5, below the pK_a of any ionizable groups in RNA. Furthermore, the small changes in substrate binding and steady-state cleavage rates from pH 8–9.5 in the RNase P RNA reaction cannot account for the plateau.[52] The plateaus in pH dependence therefore probably represent saturation of the reaction for hydroxide ion. This saturation could be kinetic, in that another step of the reaction has become rate limiting, or could represent complete conversion of the $Mg(H_2O)_n$ complex to $Mg(H_2O)_{n-1}OH^-$. In the latter interpretation, the pH at which the rate is half-maximal represents the pK_a of the Mg-aquo complex, about 8–8.5 for the RNase P RNA reaction.

Another prediction of the general base model is that the rate of the chemical step will reflect the pK_a of the metal–aquo complex in the active site. Ca^{2+}, which has a

K_a 15-fold higher than Mg, reduces the rate of the RNase P RNA reaction by 10^4.[52] Although this is consistent with a Ca^{2+} pK_a effect on the reaction, it indicates that other factors relating to metal ion identity are more important. Ca^{2+} is unable to support the *Tetrahymena* ribozyme reaction, although it appears to promote folding and substrate binding.[13,57] However, a variant of the ribozyme, produced by in vitro evolution experiments is able to use Ca^{2+} to promote catalysis.[58] The ribozymes produced by this technique are at least several hundred-fold more reactive than the starting ribozyme in the presence of Ca^{2+}. Interestingly, reactivity in the presence of Mg^{2+} is also enhanced. This result suggests that the mutated enzymes may have not acquired a new ability to bind Ca^{2+} in the active site, but are able to increase the catalytic power of any divalent cation bound.

The hypothesis that Mg-aquo complexes act as a general base in the RNase P and *Tetrahymena* ribozyme reactions is plausible and consistent with experimental data, but the evidence is not compelling. The data are also consistent with a model in which Mg^{2+} coordinate directly to the attacking water or guanosine 2'OH, and the proton is lost without going to a general base proton acceptor. There also are no data to rule out a model in which a nucleoside functional group, whose pK_a has been perturbed, is the proton acceptor.

5.4.10 Mg^{2+} as a Lewis Acid

The finding that the rate of the chemical step of the RNase P RNA reaction plateaus at high pH, rather than peaks and declines, suggests that no general acid is involved in the reaction. Further evidence for this view is provided by the characterization of a series of modifications at the substrate ribose 2' position in the *Tetrahymena* ribozyme reaction.[56] There is a very strong dependence of the rate of the chemical step of the reaction on the pK_a of the modified ribose, consistent with the absence of a general acid function. However, an acidic function is expected in order to stabilize the 3'O⁻ leaving group. Direct coordination of Mg^{2+} to the leaving group, in which Mg^{2+} acts as a Lewis acid, could provide this stabilization.

Evidence for direct coordination of Mg^{2+} to the 3'O of the substrate ribose was provided by a "Mn rescue" experiment.[40] The 3'O at the cleavage site of a deoxyribose substrate oligonucleotide was substituted by sulfur. In the presence of Mg^{2+} alone, the chemical step for cleavage of the 3'S substrate is 10^3-fold slower than for 3'O. When Mg^{2+} is supplemented with Mn^{2+}, the 3'O substrate is only cleaved eightfold faster than 3'S; when Mn^{2+} is used alone, cleavage of 3'O is only 3.5-fold faster. This 125- to 280-fold enhancement in relative rate of Mn^{2+} over Mg^{2+} is comparable to the relative affinities of Mg^{2+} and Mn^{2+} for ATP-βS (Mn^{2+}/Mg^{2+} ≈ 180).[25,26] The correlation between binding affinities and rate effects suggests that the former is largely sufficient to account for the latter.

5.4.11 Stereochemistry and Ligands

In the absence of a crystal structure, the stereochemical information we have comes from studies on the modification of substrate functional groups, such as the 3'O Mn^{2+} rescue experiment described previously. This information has been combined with model building and comparison to protein nucleases (see the following) to yield our currently incomplete view of Mg^{2+} stereochemistry in the active site.

Sugimoto et al. used the circle-opening reaction of the *Tetrahymena* IVS to study the effects of deoxyribose substitution on Mg^{2+} binding.[59] In this reaction, the terminal 3'OH of an oligonucleotide attacks the circular form of the RNA and converts it to a linear molecule[42]; this oligonucleotide corresponds to the 5' moiety of the substrate oligonucleotide, and a guanosine in the circular RNA corresponds to the free guanosine in the more familiar guanosine-dependent cleavage reaction. When the 3' terminal ribose of the circle-opening oligonucleotide is converted to deoxyribose, the binding affinity of the ribozyme–oligonucleotide complex for Mg^{2+} is weakened five- to tenfold. Mg^{2+} binding is weakened a further two- to threefold when the oligonucleotide is replaced by water as the nucleophile. Replacement of a nonterminal ribose by deoxyribose has no effect on Mg^{2+} binding, indicating that the Mg^{2+}-binding effects are specific to the terminal ribose. In the guanosine-dependent cleavage reaction, a deoxy substrate has a much higher Mg^{2+} optimum,[40] indicating that the effects of deoxy substitution on Mg^{2+} binding are not peculiar to the circle-opening reaction. These results suggest that a Mg^{2+} binding site in the active site is partially composed of the substrate ribose, and its 2'OH is especially likely to be a ligand to Mg^{2+}.

Evidence for Mg^{2+} coordination to the 3'O of the cleavage-site ribose was provided by sulfur substitution and subsequent Mn^{2+} rescue, as described.[40] In a similar experiment, the *pro*-R phosphate oxygen at the cleavage site in the *Tetrahymena* IVS reaction was replaced by sulfur.[54] In this case, the effect of thio substitution on the rate of the chemical step is small, two- to sevenfold, as compared to the 1000-fold effect noted at the 3'O. This small effect is comparable to that of thio substitution in nonenzymatic phosphate ester hydrolysis reactions in the absence of Mg^{2+}, and so supports the argument that there is no interaction of the *pro*-R_P oxygen and Mg^{2+}. Thio substitution of the *pro*-S_P oxygen has a rate effect of $\sim 10^3$-fold in the ligation reaction,[47] consistent with Mg^{2+} coordination at this site. However, rescue by Mn^{2+} addition has not been tested, and it is possible that the rate effect is due to contacts with ribozyme functional groups, or to conformational changes.

Because RNase P requires a large and properly folded substrate, chemical substitution experiments have been more difficult. Substitution by deoxyribose at the cleavage site in the RNase P RNA reaction results in a weakening of apparent binding affinity for Mg^{2+}.[17,52] This effect implies that the 2'OH acts as a ligand to Mg^{2+}, probably through water. In addition to its role as a Mg^{2+} ligand, the 2'OH probably also shares a proton with the 3' alkoxide leaving group, stabilizing its negative charge, and so facilitating its departure. Even at high concentrations of Mg^{2+}, the rate of the cleavage step is reduced 10^3-fold by 2'-deoxyribose substitution, and 10^6-fold by 2'-O-methyl substitution.[52] Coordination of Mg^{2+} to a water molecule increases the acidity of water, and therefore its ability to donate or share a proton, by 10^4-fold. A Mg^{2+}-bound water molecule could enhance the proton-sharing function of the 2'OH by donating a proton in turn to it. Similar effects of ribose 2'OH substitution in the *Tetrahymena* IVS cleavage reaction have been reported.[56]

5.4.12 Number of Active-Site Mg^{2+}

Because there are multiple Mg^{2+}-binding ligands on ribozyme substrates, and because Mg^{2+} appears to perform several functions in catalysis, it is not surprising that there are multiple Mg^{2+} in the active site. One of the intrinsic features of an RNA catalyst is the high negative charge density provided by the phosphate backbone. This property affords the opportunity for binding closely apposed divalent cations.[60]

The rate of the chemical step of cleavage by RNase P RNA shows strong cooperative dependence on Mg^{2+} concentration.[52] Kinetic analysis of the cooperativity of Mg^{2+} activation of the cleavage reaction indicates that at least three Mg^{2+} participate in catalysis.[c] In contrast, substrate binding shows no cooperativity in Mg^{2+} dependence. Deoxy substitution at the substrate phosphodiester reduces the cooperativity of Mg^{2+} activation such that the apparent number of Mg^{2+} in the reaction is reduced to two. The result could mean that only two Mg^{2+} activate the reaction, or that the cooperativity of activation of three Mg^{2+} has been reduced. This loss in cooperativity supports the hypothesis that the 2'OH at the cleavage site acts as a ligand to Mg^{2+}. Cooperativity is specific to Mg^{2+}: The Ca^{2+}-dependent reaction is only very weakly cooperative with respect to Ca^{2+} concentration.

Metal-ion cooperativity in the chemical step of the *Tetrahymena* ribozyme reaction has not yet been explicitly determined, but may be similar to RNase P RNA. Substrate cleavage rates are cooperative with respect to Mg^{2+} concentration.[13] Like RNase P RNA, at least three Mg^{2+} participate in this process. Cleavage rates were determined under conditions in which the substrate binding step is rate limiting at high (10 mM) Mg^{2+} concentrations[14]; the rate-limiting step at lower Mg^{2+} concentrations is unknown. However, substrate binding itself is not cooperative with respect to Mg^{2+} concentration,[57] and so cannot account for the observed cooperative dependence of cleavage rate. Because an excess of ribozyme over substrate was used in the cleavage reactions,[13] product release cannot be rate limiting. It is therefore plausible that cleavage rates at low Mg^{2+} concentrations are limited by the rate of the chemical step, and that the *Tetrahymena* ribozyme, like RNase P RNA, uses three Mg^{2+} in catalysis. A three-metal-ion mechanism is a common feature of protein nucleases, as will be discussed.

5.4.13 Comparison with Protein Nucleases—A Model for Metal Ion Catalysis

The crystal structures of several phosphate transfer enzymes have been analyzed with respect to metal binding at the active site, including P1 nuclease,[61] alkaline phosphatase,[62] phospholipase C,[63,64] and the exonuclease domain of *E. coli* DNA

---

[c] This number is given by the slope of the Hill plot, $\log(v/(V_{max}-v))$ vs. $\log [Mg^{2+}]$. The slope is equal to the number of ligands if cooperativity is strong—that is, all active ribozyme has all Mg sites filled. This model is referred to as the "two state" model, in which all sites are either filled or empty on a given ribozyme molecule, and no intermediate states exist. The number of ligands can be greater than the slope if cooperativity is partial.

polymerase I.[65,66] Three metal ions are found in the active sites, except for that of DNA polymerase, in which only two were detected. However, there is kinetic, spectroscopic, and NMR evidence for a third Mg^{2+},[67,68] so it is likely that DNA polymerase shares a mechanism similar to that of the other nucleases. A recent analysis of the metal sites of alkaline phosphatase, nuclease P1, and phospholipase C describes a common architecture of active-site metal binding, and a catalytic triad of metal interaction with the substrate phosphate.[69] Three zinc ions are found for phospholipase C and nuclease P1; two zinc and one Mg^{2+} are in the alkaline phosphatase active site. Two of these metal ions form a binuclear cluster, separated by 3–4 Å; the third metal is 4–7 Å distant from either of the first two. Model-building studies and analysis of cocrystals with substrate analogs indicate that all of the metal ions are able to interact with the substrate phosphate, either by direct coordination or through water.[64]

Figure 5.4 integrates the protein nuclease crystallographic data with ribozyme metal-binding data. MgA and MgB are positioned analogously to the Zn^{2+} ions in the DNA polymerase exonuclease domain.[66,70] MgB coordinates to the 3'O and *pro*-S phosphate oxygen; it acts as a Lewis acid. MgA also coordinates to the *pro*-S phosphate oxygen, as well as to the attacking nucleophile. These two ions account for the observations of Mg^{2+} coordination to the 3'O and the *pro*-S phosphate oxygen. MgC forms a hydrogen bond to the 2'OH through water, accounting for the data that suggest that this 2'OH is a ligand to a catalytically important Mg^{2+}. More speculatively, I have shown a water molecule coordinated to MgC that acts a proton acceptor from the nucleophile. All Mg^{2+} in this model share ligands in common with other Mg^{2+} ions, consistent with the observed cooperativity of metal ion activation.

5.5 Summary

Magnesium ions serve both structural and functional roles in ribozyme catalysis. The secondary structure of ribozymes, and much of its tertiary structure can form in the absence of Mg^{2+}, if other cations are present; however, Mg^{2+} is required for stabilization and optimization of tertiary structure. With the exception of the hammerhead ribozyme, RNA enzymes are highly selective in their catalytic requirements for divalent cation. There is good evidence that Mg–water complexes act as a general base in the hammerhead reaction, and by extension, other reactions in which ribose 5'O is the leaving group, that is, the delta agent and hairpin ribozymes. A single catalytic Mg^{2+} is probably sufficient to account for metal ion catalysis in these reactions.

For those reactions in which ribose 3'O is the leaving group (i.e., RNase P RNA), group I and (probably) group II self-splicing introns, the role of Mg^{2+} is more complex. There is evidence that a Mg–water complex acts as a general base, activating the nucleophile in the reaction. Mg^{2+} appears to coordinate directly to the 3'O, acting as a Lewis acid to facilitate the leaving of this group. Another site of coordination is the *pro*-S phosphate oxygen; this coordination may stabilize the

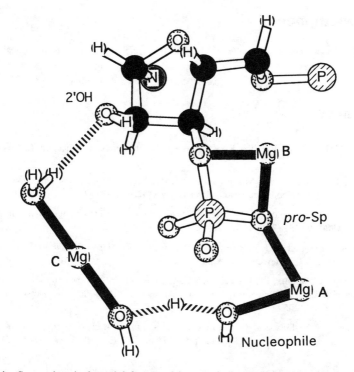

Figure 5.4 Stereochemical model for metal ion catalysis in the RNase P RNA and *Tetrahymena* IVS cleavage reactions. Covalent bonds are white, coordination bonds are solid, and hydrogen bonds are stippled. The model depicts the initiation of nucleophilic attack. The OH⁻ nucleophile shown is replaced by guanosine 2′O⁻ in the *Tetrahymena* reaction. MgA and MgB are positioned by analogy with the metal ions in the active site of *E. coli* DNA polymerase exonuclease.[66,40] Coordination to the 3′O is indicated by Mn rescue experiments.[40] Coordination to the *pro*-S_p oxygen is indicated by thio effects on cleavage rate at this position.[47,54] The presence of a third metal ion (MgC) is indicated by kinetic and spectroscopic experiments,[52,68] and by analogy with protein phosphoryl transfer enzymes.[69] Formation of a hydrogen bond contact of the 2′OH with a metal ion is indicated by substitution experiments at this position[52,56]; assignment of this contact to MgC is by default, because of the stereochemical implausibility of making such a contact with MgA or MgB. Although MgC is modeled in a position to make a coordination bond to the *pro*-R_p oxygen, the lack of a significant thio effect at this position indicates that such a bond is not made, or is not important to catalysis.[54] More speculatively, a hydroxide ion coordinated to MgC is shown acting as a general base, accepting a proton from the nucleophile. Acquisition of this proton renders the other water molecule shown coordinated to MgC more able to share a proton with the 2′OH, in turn enhancing its ability to stabilize the negative charge of the 3′O leaving group.

trigonal bipyramidal transition-state geometry. The 2′OH appears to bind Mg^{2+}, probably through water. This interaction may serve to enhance proton sharing between 2′OH and 3′O, further facilitating departure of the leaving group. Kinetic analysis and comparison with protein nucleases indicates a three-metal-ion mechanism.

Acknowledgments

I thank Drs. Olke Uhlenbeck, Michael Yarus, Norman Pace, and Torsten Wiegand for critical review and helpful comments.

References

1. Yarus, M., How Many Catalytic RNAs? Ions and the Cheshire Cat Conjecture, *FASEB J.* **1993**, *7*, 31–39.

2. Labuda, D.; Porschke, D., Magnesium ion inner sphere complex in the anticodon loop of phenylalanine transfer ribonucleic acid, *Biochemistry* **1982**, *21*, 49–53.

3. Schimmel, P.; Redfield, A., Transfer RNA in solution: selected topics, *Annu. Rev. Biophys. Bioeng.* **1980**, *9*, 181–221.

4. Teeter, M. M.; Quigley, G. J.; Rich, A., in *Nucleic acid–metal ion interactions* (Spiro, T. G., Ed.), **1980**, pp. 147–77, Wiley, New York.

5. Heerschap, A.; Walters, J.A.L.I.; Hilbers, C. W., Interactions of some naturally occurring cations with phenylalanine and initiator tRNA from yeast as reflected by their thermal stability, *Biophys. Chem.* **1985**, *22*, 205–17.

6. Sampson, J. R.; Uhlenbeck, O. C., Biochemical and physical characterization of an unmodified yeast phenylalanine transfer RNA transcribed in vitro, *Proc. Natl. Acad. Sci. U.S.A* **1988**, *85*, 1033–37.

7. Romer, R.; Hach, R., tRNA conformation and magnesium binding, *Eur. J. Biochem.* **1975**, *55*, 271–84.

8. Bolton, P. H.; Kearns, D. R., Effect of cations on tRNA structure, *Biochem.* **1977**, *16*, 5729–41.

9. Bujalowski, W.; Graeser, E.; McLaughlin, L. W.; Porschke, D., Anticodon loop of tRNA^{Phe}: Structure, dynamics, and Mg^{2+} Binding, *Biochemistry* **1986**, *25*, 6365–71.

10. Labuda, D.; Nicoghosian, K.; Cedergren, R., Cooperativity in low-affinity Mg^{2+} binding to tRNA, *J. Biol. Chem.* **1985**, *260*, 1103–7.

11. Labuda, D.; Haertle, T.; Augustyniak, J., Dependence of tRNA structure in solution upon ionic condition of the solvent, *Eur. J. Biochem.* **1977**, *79*, 293–301.

12. Jaeger, J.; Zuker, M.; Turner, D., Melting and chemical modification of a cyclized self-splicing Group I intron: Similarity of structures in 1 M Na$^+$, in 10 mM Mg^{2+}, and in the presence of substrate, *Biochemistry* **1990**, *29*, 10147–58.

13. Celander, D. W.; Cech, T. R., Visualizing the higher order folding of a catalytic RNA molecule, *Science* **1991**, *251*, 401–7.

14. Herschlag, D.; Cech, T. R., Catalysis of RNA cleavage by the *Tetrahymena thermophila* ribozyme. 1. Kinetic description of the reaction of an RNA substrate complementary to the active site, *Biochemistry* **1990**, *29*, 10159–71.

15. Burgin, A. B.; Pace, N. R., Mapping the active site of ribonuclease P RNA using a substrate containing a photoaffinity agent, *EMBO J.* **1990**, *9*, 4111–18.

16. Smith, D.; Burgin, A. B.; Haas, E. S.; Pace, N. R., Influence of metal ions on the ribonuclease P reaction. Distinguishing substrate binding from catalysis, *J. Biol. Chem.* **1992**, *267*, 2429–36.

17. Perreault, J.; Altman, S., Pathway of activation by magnesium ions of substrates for the catalytic subunit of RNAse P from Escherichia coli, *J. Mol. Biol.* **1993**, *230*, 750–56.

18. Chowrira, B.; Burke, J., Extensive phosphorothioate substitution yields highly active and nuclease-resistant hairpin ribozymes, *Nucleic Acids Res.* **1992**, *20*, 2835–40.

19. Heus, H.; Pardi, A., Nuclear magnetic resonance studies of the hammerhead ribozyme domain. Secondary structure formation and magnesium ion dependence, *J. Mol. Biol.* **1991**, *217*, 113–24.

0. Koizumi, M.; Ohtsuka, E., Effects of phosphorothioate and 2-amino groups in hammerhead ribozymes on cleavage rates and Mg^{2+} binding, *Biochemistry* **1991**, *30*, 5145–50.

21. Holbrook, S. R.; Sussman, J. L.; Warrant, R. W.; Church, G. M., RNA–ligand interactions (I) magnesium binding sites in yeast tRNA[Phe], *Nucl. Acids Res.* **1977**, *4*, 2811–19.

22. Jack, A.; Ladner, J. E.; Rhodes, D.; Brown, R. S.; Klug, A., A crystallographic study of metal-binding to yeast phenylalanine transfer RNA, *J. Mol. Biol.* **1977**, *111*, 315–28.

23. Quigley, G. J.; Teeter, M. M.; Rich, A., Structural analysis of spermine and magnesium ion binding to yeast phenylalanine transfer RNA, *Proc. Natl. Acad. Sci. USA* **1978**, *75*, 64–68.

24. Tu, A. T.; Heller, M. J.; Structure and stability of metal–nucleoside phosphate complexes, in *Metal Ions in Biological Systems: Simple Complexes* (H. Sigel, Ed.) Marcel Dekker, New York, **1974**, 1–49.

25. Jaffe, E.; Cohn, M., Diastereomers of the nucleoside phosphorothioates as probes of the structure of the metal nucleotide substrates and of the nucleotide binding site of yeast hexokinase, *J. Biol. Chem.* **1979**, *254*, 10839–45.

26. Pecoraro, V.; Hermes J.; Cleland, W., Stability constants of Mg^{2+} and Cd^{2+} complexes of adenine nucleotides and thionucleotides and rate constants for formation and dissociation of MgATP and MgADP, *Biochemistry* **1984**, *23*, 5262–71.

27. Christian, E. L.; Yarus, M., Metal coordination sites that contribute to structure and catalysis in the group I intron from Tetrahymena, *Biochemistry* **1993**, *32*, 4475–80.

28. Michel, F.; Westhof, E., Modelling of the three-dimensional architecture of group I catalytic introns based on comparative sequence analysis, *J. Mol. Biol.* **1990**, *216*, 585–610.

29. Werner, C.; Krebs, B.; Keith, G.; Dirheimer, G., Specific cleavages of pure tRNAs by plumbous ions, *Biochim. Biophys. Acta* **1976**, *432*, 161–75.

30. Brown, R.; Hingerty, B.; Dewan, J.; Klug, A., Pb(II)-catalysed cleavage of the sugar–phosphate backbone of yeast tRNAPhe—Implications for lead toxicity and self-splicing RNA, *Nature* **1983**, *303*, 543–46.

31. Brown, R.; Dewan, J.; Klug, A., Crystallographic and biochemical investigation of the lead(II)-catalyzed hydrolysis of yeast phenylalanine tRNA, *Biochemistry* **1985**, *24*, 4785–801.

32. Behlen, L.; Sampson, J.; DiRenzo, A.; Uhlenbeck, O., Lead-catalyzed cleavage of yeast tRNA[Phe] mutants, *Biochemistry* **1990**, *29*, 2515–23.

33. Streicher, B.; von Ahsen, U.; Shroeder, R., Lead cleavage sites in the core structure of group I intron-RNA, *Nucl. Acids Res.* **1993**, *21*, 311–17.

34. Kazakov, S.; Altman, S., Site-specific cleavage by metal ion cofactors and inhibitors of M1 RNA, the catalytic subunit of RNAse P from Escherichia coli, *Proc. Natl. Acad. Sci. USA*, **1991**, *88*, 9193–97.

35. Brown, J. W.; Pace, N. R., Ribonuclease P RNA and protein subunits from bacteria, *Nucl. Acids Res.*, **1992**, *20*, 1451–56.

36. Dahm, S.; Uhlenbeck, O., Role of divalent metal ions in the hammerhead RNA cleavage reaction, *Biochemistry* **1991**, *30*, 9464–69.

37. Chowrira, B.; Berzal-Herranz, A.; Burke, J., Ionic requirements for RNA binding, cleavage, and ligation by the hairpin ribozyme, *Biochemistry* **1993**, *32*, 1088–95.

38. Noller, H.; Hoffarth, V.; Zimniak, L., Unusual resistance of peptydyl transferase to protein extraction procedures, *Science* **1992**, *256*, 1416–19.

39. Piccirilli, J.; McConnell, T.; Zaug, A.; Noller, H.; Cech, T., Aminoacyl esterase activity of the Tetrahymena ribozyme, *Science* **1992**, *256*, 1420–24.

40. Piccirilli, J.; Vyle, J.; Caruthers, M.; Cech, T., Metal ion catalysis in the Tetrahymena ribozyme reaction, *Nature* **1993**, *361*, 85–88.

41. Haydock, K.; Allen, L. C., Molecular mechanism of catalysis by RNA, *Prog. in Clinical and Biological Research* **1985**, 87–98, A. R. Liss, Inc., New York.

42. Cech, T. R., The chemistry of self-splicing RNA and RNA enzymes, *Science* **1987**, *236*, 1532–39.

43. Gardiner, K. J.; Marsh, T. L.; Pace, N. R., Ion dependence of the *Bacillus subtilis* RNase P reaction, *J. Biol. Chem.* **1985**, *260*, 5415–19.

44. Guerrier-Takada, C.; Haydock, K.; Allen, L.; Altman, S., Metal ion requirements and other aspects of the reaction catalyzed by M1 RNA, the RNA subunit of ribonuclease P from *Escherichia coli*, *Biochemistry* **1986**, *25*, 1509–15.

45. Grosshans, C.; Cech, T., Metal ion requirements for sequence-specific endoribonuclease activity of the Tetrahymena ribozyme, *Biochemistry* **1989**, *28*, 6888–94.

46. Hardt, W.; Schlegl, J.; Erdmann, V.; Hartmann, R., Gel retardation analysis of E. coli M1 RNA–tRNA complexes, *Nucleic Acids Res.* **1993**, *21*, 3521–27.

47. Rajagopal, J.; Doudna, J.; Szostak, J., Stereochemical course of catalysis by the Tetrahymena ribozyme, *Science* **1989**, *244*, 692–94.

48. McSwiggen, J.; Cech, T., Stereochemistry of RNA cleavage by the Tetrahymena ribozyme and evidence that the chemical step is not rate-limiting, *Science* **1989**, *244*, 679–83.

49. van Tol, H.; Buzayan, J.; Feldstein, P.; Eckstein, F.; Bruening, G., Two autolytic processing reactions of a satellite RNA proceed with inversion of configuration, *Nucleic Acids Res.* **1990**, *18*, 1871–75.

50. Fedor, M.; Uhlenbeck, O., Kinetics of intermolecular cleavage by hammerhead ribozymes, *Biochemistry* **1992**, *31*, 12042–54.

51. Reich, C. I.; Olsen, G. J.; Pace, B.; Pace, N. R., Role of the protein moiety of ribonuclease P, a ribonucleoprotein enzyme, *Science* **1988**, *239*, 81.

52. Smith, D.; Pace, N., Multiple magnesium ions in the ribonuclease P reaction mechanism, *Biochemistry* **1993**, *32*, 5273–81.

53. Pan, T.; Long, D.; Uhlenbeck, O., Divalent metal ions in RNA folding and catalysis, Cold Spring Harbor Laboratory, Cold Spring Harbor (R. Gesteland, R.; Atkins, J., Eds.) *The RNA World*, 1993, pp. 271–302.

54. Herschlag, D.; Piccirilli, J.; Cech, T., Ribozyme-catalyzed and nonenzymatic reactions of phosphate diesters: Rate effects upon substitution of sulfur for a nonbridging phosphoryl oxygen atom, *Biochemistry* **1991**, *39*, 4844–54.

55. Izatt, R. M.; Hansen, L. D.; Rytting, J. H.; Christensen, J. J., Thermodynamics of proton dissociation in dilute aqueous solution. V. An entropy titration study of adenosine, pentoses, hexoses, and related compounds, *J. Am. Chem. Soc.* **1966**, *88*, 2641–45.

56. Herschlag, H.; Eckstein, F.; Cech, T., The importance of being ribose at the cleavage site in the *Tetrahymena* ribozyme reaction, *Biochemistry* **1993**, *32*, 8312–21.

57. Pyle, A.; McSwiggen, J.; Cech, T., Direct measurement of oligonucleotide substrate binding to wild-type and mutant ribozymes from Tetrahymena, *Proc. Natl. Acad. Sci. U.S.A* **1990**, *87*, 8187–91.

58. Lehman, N.; Joyce, G., Evolution *in vitro* of an RNA enzyme with altered metal dependence, *Nature* **1993**, *361*, 182–85.

59. Sugimoto, N.; Tomka, M.; Kierzek, R.; Bevilacqua, P. C.; Turner, D. C., Effects of substrate structure on the kinetics of circle opening reactions of the self-splicing intervening sequence from *Tetrahymena thermophila:* Evidence for substrate and Mg^{2+} binding interactions, *Nucl. Acids. Res.* **1989**, *17*, 355–71.

60. Pace, N. R.; Smith, D., Ribonuclease P: Function and variation, *J. Biol. Chem.* **1990**, *265*, 3587.

61. Volbeda, A.; Lahm, A.; Sakiyama, F.; Suck, D., Crystal structure of *Penicillium citrinum* P1 nuclease at 2.8 Å resolution, *EMBO J.* **1991**, *10*, 1607–18.

62. Sowadski, J. M.; Handschmacher, M. D.; Krishna Murthy, H. M., Foster, B. A.; Wyckoff, H. W., Refined structure of alkaline phosphatase from *Escherichia coli* at 2.8 Å resolution, *J. Mol. Biol.* **1985**, *186*, 417–33.

63. Hough, E.; Hansen, L. K.; Birknes, B.; Jynge, K.; Hansen, S.; Hordvik, A.; Little, C.; Dodson, E.; Derewenda, Z., High-resolution (1.5 Å) crystal structure of phopholipase C from *Bacillus cereus, Nature* **1989**, *338*, 357–60.

64. Byberg, J.; Jorgensen, F.; Hansen, S.; Hough, E., Substrate–enzyme interactions and catalytic mechanism in phospholipase C: A molecular modeling study using the GRID program, *Proteins* **1992**, *12*, 331–38.

65. Freemont, P.; Friedman, J.; Sanderson, M.; Steitz, T., Cocrystal structure of an editing complex of Klenow fragment with DNA, *Proc. Natl. Acad. Sci. USA* **1988**, *85*, 8924–28.

66. Beese, L. S.; Steitz, T. A., Structural basis for the 3'-5' exonuclease activity of the *Escherichia coli* DNA polymerase I: A two metal ion mechanism, *EMBO J.* **1991**, *10*, 25–33.

67. Mullen, G.; Serpersu, E.; Ferrin, L.; Loeb, L.; Mildvan, A., Metal binding to DNA polymerase I, its large fragment, and two 3',5'-exonuclease mutants of the large fragment, *J. Biol. Chem.* **1990**, *265*, 14327–34.

68. Han, H.; Rifkind, J. M.; Mildvan, A. S., Role of divalent cations in the 3',5'-exonuclease reaction of DNA polymerase I, *Biochemistry* **1991**, *30*, 11104–8.

69. Vallee, B.; Auld, D., New perspective on zinc biochemistry: Cocatalytic sites in multi-zinc enzymes, *Biochemistry* **1993**, *32*, 6493–500.

70. Steitz, T.; Steitz, J., A general two-metal-ion mechanism for catalytic RNA, *Proc. Natl. Acad. Sci. USA* **1993**, *90*, 6498–502.

71. Surratt, C. K.; Carter, B. J.; Payne, R. C.; Hecht, S. M., Metal ion and substrate structure dependence of the processing of transfer RNA precursors by RNase P and M1 RNA, *J. Biol. Chem.* **1990**, *265*, 22513–19.

72. Wu, H.; Lin, Y.; Lin, F.; Makino, S.; Chang, M.; Lai, M., Human hepatitis delta virus RNA subfragments contain an autocleavage activity, *Proc. Natl. Acad. Sci. USA* **1989**, *86*, 1831–35.

73. Rosenstein, S.; Been, M., Self-cleavage of hepatitis delta virus genomic strand RNA is enhanced under partially denaturing conditions, *Biochemistry* **1990**, *29*, 8011–16.

74. Summers, D., *Chemistry Handbook,* 1975, Willard Grant Press, Boston.

75. Burgess, J., *Ions in Solution: Basic Principles of Chemical Interactions,* Ellis Horwood Ltd., Chichester, England, 1988.

6

Magnesium-Dependent Enzymes in Nucleic Acid Biochemistry

C. B. Black and J. A. Cowan

6.1 Introduction

Enzymes that catalyze the hydrolysis and formation of phosphodiester bonds underlie the processing of genetic information within the cell. Such pathways include the transcription, translation, and replication of nucleic acids, and almost all of the enzymes utilized in these reactions require divalent magnesium as an essential cofactor for optimal activity. Nevertheless there is currently no detailed mechanistic understanding of the functional role of magnesium ion as an enzyme activator, although this issue is now the focus of intense investigation.

In this chapter we illustrate the chemistry of magnesium as it relates to a representative group of enzymes that have been most widely studied. This is by no means a comprehensive review, reflecting more the paucity of data on this subject, but several general trends do emerge that provide basic mechanistic insight on these reaction pathways.

6.2 Polymerase Reactions

Polymerization of genetic material occurs through the synthesis of phosphodiester bonds from single nucleotide building blocks and employs numerous enzymes that are required to catalyze phosphodiester bond formation. Magnesium ion serves a variety of roles that include mediating the interaction of enzyme with substrate, allosteric control of enzyme structure, and ATPase or phosphoryl transfer activity.[a] For example, RNA polymerase uses magnesium to effect an allosteric change in protein structure and to drive catalysis through ATP hydrolysis.[1] In this section two polymerase enzymes are reviewed that best illustrate the current thinking on this enzyme class, and that also illustrate the various functional roles for magnesium ion.

6.2.1 DNA Polymerase I (Polymerase Site)

DNA polymerase I is a multifunctional enzyme that utilizes magnesium and zinc as essential metal cofactors. Pol I displays 5'-3' polymerase, 3'-5' exonuclease, and 5'-3' exonuclease activities. The protein can be proteolytically cleaved into a 68 kDa fragment (Klenow fragment) that contains the polymerase and 3'-5' exonuclease functions, and a smaller 35 kDa fragment that contains the 5'-3' exonuclease activity. The 3'-5' exonuclease site will be discussed later.

To stimulate maximal activity the concentration of Mg^{2+} must lie in the range of 5–10 mM.[2] By use of Mn^{2+} as a probe ion, EPR studies have revealed one binding site for a divalent cation in the polymerase active site, with a $K_d(Mn^{2+}) = 3.6$ mM.[3] This site is generated by the simultaneous binding of dGTP to form a ternary complex.[3] In addition to EPR studies, proton resonance relaxation investigations show at least one water molecule bound to the manganese ion.[b] With two or three oxygens from dGTP, and one bound water molecule, the two or three remaining ligands must come from the enzyme.[3] A competition study using the polymerase–dGTP complex with both Mn^{2+} and Mg^{2+} gave a dissociation constant for magnesium $K_d(Mg^{2+}) = 100 \pm 2$ μM. Mutagenesis of the proposed active-site residues (Arg-668, Asp-882, and Gln-849) resulted in a drop in k_{cat} to almost zero when any *one* of the residues was mutated (Table 6.1), and so a catalytic role has been ascribed to each of the three residues.[4] Clearly, however, there is also a change in the binding affinity of the DNA substrate (K_d) in a manner that does not correlate in a meaningful way with the charge on the side chain. It is not yet clear which if any of these side chains or backbone carbonyls are involved in binding of the metal cofactor. Two roles have been proposed for this cofactor. First, magnesium serves a structural role by binding to the single-strand nucleotide and stabilizing a conformation required either for catalysis, protein recognition, or both.[5–7] Second, kinetic data support the formation of a metal binding site after nucleotide addition, and so it has been proposed that the metal ion organizes the active site by bridging nucleotide

---

[a] Abbreviations used: ADP, adenosine diphosphate; ATP, adenosine triphosphate; DNA, deoxyribonucleic acid; Da, Dalton; EPR, electron paramagnetic resonance; GTP, guanosine triphosphate; NAD⁺, nicotinamide adenine dinucleotide; NMR, nuclear magnetic resonance; PEP, phosphoenolpyruvate; RNA, ribonucleic acid; RNAP, RNA polymerase.

[b] See Chapter 3 for a more complete discussion of the application of physical methods to monitor the binding chemistry of transition metal ion probes.

Table 6.1 STEADY-STATE KINETIC PARAMETERS AND SUBSTRATE
DISSOCIATION CONSTANTS (K_d) OBTAINED FOR
MUTANTS OF THE *E. coli* DNA POLYMERASE I
(KLENOW FRAGMENT) POLYMERASE ACTIVITY[a]

Mutation	K_m(dNTP) (μM)	k_{cat} (s^{-1})	K_d (nM)
Wild type	2.3	2.8	8
R668A	6.5	0.006	0.5
Q849A	3.8	0.02	100
D882A	4.1	0.006	140

[a] Steady-state data were obtained with 2 mM Mg^{2+}. The substrate dissociation constant K_d was determined using ^{32}P-labeled DNA in 5 mM Mg^{2+} solution.[4]

contacts and enzyme contacts,[3,4] Evidence for both hypotheses is sparse, and neither has been conclusively demonstrated.

6.2.2 RNA Polymerase

Translation of proteins ultimately begins with the transcription of genomic DNA to mRNA. The enzyme directly responsible for the conversion of DNA to RNA is RNA polymerase (RNAP). In prokaryotes there is only one enzyme to carry out this function; however, in eukaryotes there are three types of RNA polymerase that function according to specific classes of genes. RNA polymerase I (RNAP I) transcribes class I genes that code for ribosomal RNA; RNAP II transcribes genes that code for proteins; and RNAP III transcribes genes that encode small RNA molecules.

Mildvan and co-workers have reported seven binding sites for divalent cations by Mn^{2+} EPR experiments, including a unique "tight" site and several "weak" sites. The tight-binding site (K_d = 1.9 μM) is believed to correspond to the catalytic site, while the location of the weak-binding ions are unknown ($K_{d_{app}}$ = 200 μM).[8]

In vitro studies have shown that the extent of the processivity and conformational flexibility of RNAP is dependent upon magnesium ion concentration.[9] A three-step mechanism has been proposed, with two conformational changes prior to the catalytic step (Scheme 6.1), where R = RNA and P = RNAP. Note how the magnesium ions aid in the conformational switch from the closed to the open form.

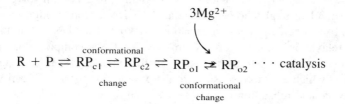

$$3Mg^{2+}$$

$$R + P \rightleftharpoons RP_{c1} \rightleftharpoons RP_{c2} \rightleftharpoons RP_{o1} \rightleftharpoons RP_{o2} \cdots catalysis$$

conformational change   conformational change

Scheme 6.1

After intial substrate binding, the enzyme adopts the closed position (RP_{c1}) and subsequently undergoes a kinetically distinct change to a second closed (RP_{c2}) form or intermediate. Step three involves a change to an open form where the DNA strands are partially melted over an 11–16 base pair region.[10] It is thought that Mg^{2+} effects this conformational change to a catalytically competent state. The two open forms are observed only when magnesium ions are present.[11] By evaluating the rate of association (k_a) of RNAP and its promoter substrate, and the dissociation rate (k_d) of the open complex, a requirement for uptake of approximately 3 Mg^{2+} was deduced.[11] The structural mechanisms whereby magnesium ions induce these conformational transitions remain unclear. Three possibilities include allosteric binding to the enzyme, stabilization of the bound substrate that induces a new k_d, or a combination of both. It remains uncertain whether the metal cofactors interact predominantly with the RNA substrate or protein side chains.

6.3 Regulation of Topology

The chemistry of nucleic acids is often dependent on enzymes that regulate the topology (tertiary or quarternary structure) of polynucleotides. One class of enzymes that are designed to interconvert tertiary structures are the topoisomerases. The activity of such enzymes is often linked to ATP hydrolysis.

6.3.1 Topoisomerase I

Eukaryotic and prokaryotic type I topoisomerases are metalloproteins that serve to alter the topology of circular polynucleotides by transient cleavage of one strand (in a single-stranded region), followed by intrastrand passage and subsequent ligation. This class of enzyme is believed to prime nucleic acids for polymerization or modification reactions by other enzymes.[12,13] Several reactions are mediated by topoisomerase, including relaxation of the DNA substrate by strand scission and strand crossing, followed by ligation. Initially the enzyme recognizes and binds to a target site, forming a covalent complex between an active site tyrosine and one of the strands (Fig. 6.1).[14] Although formation of this complex does not require Mg^{2+}, relaxation of the ds-DNA substrate does require the metal cofactor. In addition, processivity is regulated by Mg^{2+}; increasing divalent cation concentration increases the degree of processivity of the enzyme, while increased monovalent cation shows an increase in distributive behavior.[15] After binding, relaxation involves the passing of one strand through another to "relax" the DNA from a high-energy supercoiled form to an open circular conformation. Nicking of the DNA strand is a prerequisite for relaxation. Since hydrolysis of the Tyr–DNA complex does not occur with ds-DNA as the substrate, it is thought that the enzyme nicks single-stranded regions,[16] which are more sensitive to the presence or absence of cations.

Both the cleavage and relaxation activities have been found to require Mg^{2+} in prokaryotic cells only. Eukaryotic cells can function independently of magnesium,[15]

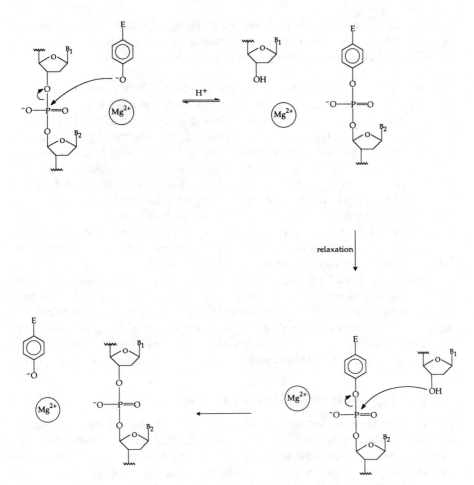

Figure 6.1 Putative reaction scheme for topoisomerase I activity showing an active-site tyrosine and magnesium. The first bond cleavage step results in relaxation of the coiled DNA via a strand passage event. Step two involves religation of the broken strand. The role of magnesium is not entirely clear, although it may involve electrostatic stabilization of the tyrosinate anion.[16,18]

although maximal activity is different. For the cleavage reaction, maximal activity requires 15 mM Mg^{2+} (it increases linearly from 2 mM), while maximal rates for strand relaxation require 2.5 mM Mg^{2+}.[16] Since the final ligation involves nucleophilic attack from a 3′-OH on the ribose ring, it has been suggested that the concentration of Mg^{2+} needed to stabilize reaction intermediates is much less than the concentration required to generate a nucleophilic hydroxyl anion for the cleavage reaction. Inhibition kinetics for relaxation were not observed in cases where magnesium ion concentration was varied between 15–20 mM.[15,16] However, Mg^{2+}

concentrations exceeding 20 mM $MgCl_2$ do show inhibition of initial velocity measurements.[15] This upper limit in concentration was extended to 60 mM by use of $Mg(aspartate)_2$, suggesting a role for the Cl^- ions in the mechanism of inhibition.

Current thinking points to two reasonable possibilities for the role of magnesium in topoisomerase I function. First, the bound metal cofactor orchestrates structural changes in the active site that promote trapping of the supercoiled DNA. In this model the cofactor binds predominantly to the enzyme. Alternatively, the metal ion can bind to the supercoiled DNA (presumably in the presence of the enzyme) and change its structure and/or dynamics to stabilize the enzyme–substrate complex.[17] In terms of catalysis it has been suggested that magnesium ion does not act as a Lewis acid by direct coordination to the phosphate backbone, inasmuch as the substitutionally inert complex $Co(NH_3)_6^{3+}$ activates topoisomerase I toward relaxation.[18] A postulate based on these data is that the cofactor enhances ionization of an active-site tyrosine to form the nucleophilic tyrosinate anion, which attacks the DNA backbone and forms a transient covalent intermediate.

Various other roles for the divalent cofactor have also been proposed. For example, the cation(s) assists with orientation of the single-stranded nucleotide chain so that backbone scission may proceed. Alternatively, the cation may regulate catalysis at the hydrolysis step (as seen from the positive correlation between cleavage and magnesium concentration). A third role portrays the enzyme-bound cofactor assisting in strand passage by mediating protein–DNA conformational changes.[16]

6.3.2 Topoisomerase II (DNA Gyrase)

Type II topoisomerases (or DNA gyrase in bacterial systems) act on eukaryotes (or bacteria) by altering the topology of DNA by passing an intact helix through a transient, enzyme-bound, double-stranded break made in the second helix. In prokaryotic systems, gyrase functions to catalyze the negative supercoiling of covalently closed DNA. In both cases the catalytic steps are coupled to ATP hydrolysis. For gyrase activity, the reaction sequence is intiated by enzyme binding to substrate DNA with a segment of the DNA wrapping around the protein (in a positive sense.).[19] This enzyme-bound section is cleaved in both strands with concomitant formation of covalent bonds between the 5′-terminae and tyrosines in the catalytic subunit of the enzyme. The remaining segment of DNA is passed through this break and then resealed.[19] Topoisomerase II shows evidence of two distinct magnesium binding sites, one of which directly participates in the cleavage reaction,[20] while a second is involved in the enzyme-mediated ATPase reaction.[21]

In prokaryotes (E. coli), the enzyme is made up of two subunits (A_2B_2), having molecular masses of 97 and 90 kDa, respectively. The enzyme binds to DNA by forming a complex that wraps around the protein core and covers approximately 120 base pairs. Only recently has a high-resolution structure (2.5 Å) become available for the B subunit.[22] The A subunit, for which the structure is not known, is believed to catalyze the relaxation mechanism and requires Mg^{2+} for activity.[23] A two-dimensional structure has been solved using a lipid monolayer.[24] While a number of mechanistic studies have supported the conclusion that subunit A is associated with the cleavage and ligation of DNA, subunit B appears to catalyze ATP hydrolysis.[19]

Studies relating to the role of the magnesium cofactor have thus far been limited to the eukaryotic systems. In particular, the dependence of reaction velocity on magnesium concentration has been investigated by monitoring the percent cleavage reaction and ATPase activity. For the cleavage reaction, the half-saturation constants (the magnesium concentration required for half-maximal activity) is ~ 2 mM (Fig. 6.2).[25] Also, inhibition is observed at magnesium concentrations above ~ 8 mM ($K_i \sim 11$ mM; the concentration of magnesium for half-maximum inhibition or half of the normal activity).[25] The ATPase reaction yielded $K_{\frac{1}{2}} \sim 0.75$ mM with an optimal magnesium concentration of 1.8 mM.

In eukaryotic DNA gyrase, Mg^{2+} is likely to play two principal roles. First, participating directly in enzyme-mediated hydrolytic cleavage of DNA, and second, promoting ATPase activity for phosphoryl transfer.[25,27] Although Mg^{2+} appears to be important in the cleavage reaction (which occurs prior to ATP hydrolysis), it was not found to be required for recognition of DNA topology, selection of cleavage sites, quarternary structure formation (the homodimer is the active form), or formation of the enzyme–DNA complex (where it is 75% bound).[25]

On the basis of recent crystallographic evidence, phosphorylation of subunit B by ATP is believed to facilitate the large structural changes in the protein necessary for strand passage after DNA cleavage. The energy derived from ATP binding to the B monomer is likely to stabilize an energetically unfavorable conformation prior to strand passage. After hydrolysis, this energy is released, and the protein returns to the relaxed state.[22] Inasmuch as the extent of ATP hydrolysis is proportional to the

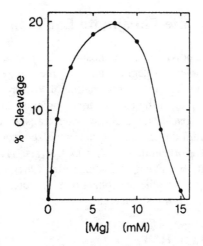

Figure 6.2 Kinetic data for the topoisomerase II (or DNA gyrase) cleavage reaction, showing the variation of the initial velocity with Mg^{2+} concentration. Activity increases until ca. 8 mM Mg^{2+}, and then decreases as magnesium ion becomes inhibitory. The inhibition mechanism is currently unknown. Adapted from reference 25.

amount of Mg^{2+}-ATP present,[25] it has been concluded that Mg^{2+} interacts with the *B* subunit solely as the Mg^{2+}-ATP–substrate complex.

6.3.3 Helicases

DNA helicases were first isolated in 1976 from an *E. coli* strain.[28] This class of enzyme usually binds to one strand of a substrate DNA duplex and moves unidirectionally, simultaneously unwinding the opposite strand.[29] This is a required step prior to the action of other DNA processing enzymes (such as polymerases). Several proposals have been made to rationalize the utility of helicase function. In particular, it is observed that helicases generate single-stranded regions for recognition by other enzymes, thereby increasing the specificity for a defined site on the nucleotide backbone. Helicases also provide a guide mechanism that targets a specific site for modification reactions. Moreover, it serves to protect other reaction sites from modification, polymerization, or hydrolytic cleavage. Related enzymes for RNA–RNA and RNA–DNA have also been identified in transcription, translation, and RNA splicing.[30-32] Other than forming a substrate complex with ATP, specific roles for Mg^{2+} are difficult to define. It is known that low magnesium concentrations result in disruption of the helix structure (and consequently protein specificity), and so a role in preserving substrate conformation has been implicated.[33] Unfortunately, the understanding of this enzyme class is complicated by the immense diversity in properties and numbers of helicases (for example, there are 10 types in *E. coli* alone).[34] Kinetic studies demonstrate that magnesium influences the unwinding reaction with a $K_{\frac{1}{2}} \sim 0.8$ mM, while maximal activity is reached at ~ 2 mM $[Mg^{2+}]$.[35]

6.4 Hydrolysis of the Phosphate Backbone

Hydrolytic cleavage of a polynucleotide phosphate backbone is ubiquitous in nucleic acid biochemistry. Nucleotide degradation, polymerase editing mechanisms, and topological changes in nucleic acid conformation all require at least transient backbone scission. For example, in the case of topoisomerase discussed earlier, a nick must first be made in the backbone to allow strand transfer. Hydrolysis is probably the best known magnesium-dependent enzyme-catalyzed reaction, in part as a result of the large amount of work done on ATP hydrolysis. Hydrolysis of phosphate esters typically proceeds by attack of nucleophilic water or hydroxide. Divalent magnesium may act to polarize the phosphate bonds, increase the acidity of Mg^{2+}-bound waters, or stabilize a transition state.

6.4.1 Ribonuclease H

Ribonuclease H (RNase H) is a nonspecific endonuclease that hydrolytically cleaves the RNA strand of RNA·DNA hybrids. The enzyme is made up of a single polypeptide chain, containing 155 amino acids, and is functionally active as a monomer.

The in vivo function of the protein has not been completely determined; however, its main function appears to be the hydrolysis of primers or intermediates during transcription.

Crystal structures are available for both the *E. coli* enzyme and the RNase H domain of HIV-I reverse transcriptase (HIV I RT).[36,37] Both crystallographic data and mutagenesis studies of the *E. coli* enzyme show that the magnesium ion binds in a pocket containing three carboxylate residues: Asp-10, Glu-48, and Asp-70 (*E. coli*). For the *E. coli* enzyme it is thought that substrate recognition occurs through both electrostatic interactions in the minor groove and hydrogen bonding to the 2′-OH of the RNA strand (Fig. 6.3).[38–40] Currently there is no evidence for a substantive role for Mg^{2+} in substrate recognition.

Characterization of the magnesium site by ^{25}Mg NMR methods has yielded on–off rates for magnesium binding: $k_{on} \sim 3.1 \times 10^8$ s^{-1}, $k_{off} \sim 1.1 \times 10^4$ s^{-1}. An association constant $K_a \sim 10^4$ M^{-1} has been determined through titration micro-calorimetry experiments (Fig. 6.4) and ^{25}Mg NMR, with no evidence of secondary sites.[42] One method that has been developed to probe the catalytically active cation is the use of substitutionally inert probe ions, such as $Co(NH_3)_6^{3+}$.[43,44] Experimental results with such probe ions suggest that the active magnesium cofactor does not serve as a Lewis acid by direct coordination when activating the phosphodiester substrate to hydrolysis, and most likely acts in an outer-sphere fashion. One feasible mode of action for Mg^{2+} is to stabilize intermediates through electrostatic interac-

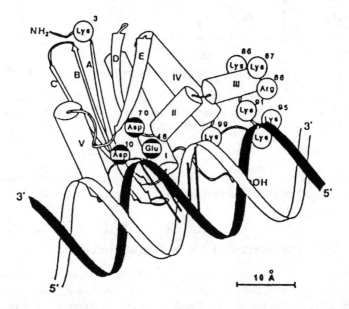

Figure 6.3 Schematic illustration of *E. coli* ribonuclease H with the handle region and active-site residues shown. Also, the hybrid helix has been modeled and superimposed on the structure to indicate a possible substrate binding mode (black indicates RNA strand, white is DNA). Adapted from reference 39.

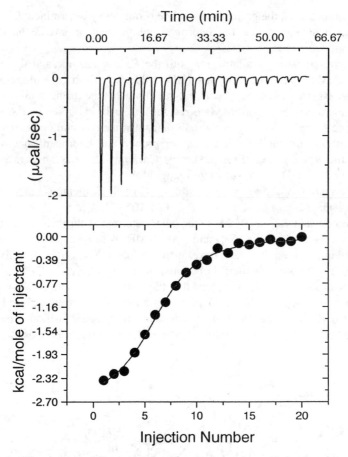

Figure 6.4   Titration calorimetry data for *E. coli* ribonuclease H showing heat output as a function of added Mg^{2+}. The data demonstrate one divalent cation binding site on the protein with a binding constant of ca. 200 μM.[42]

tion and/or hydrogen bond formation. Efforts to quantify such interactions by ^{1}H NMR and [59]Co NMR show that both the hydrogen bonding and electrostatic components are significant.[45]

6.4.2 DNA Polymerase I (3′-5′ Exonuclease Site)

DNA polymerase I is a replication and repair enzyme that shows three distinct activities. Two of these, polymerase and 3′-5′ exonuclease activity, are located on the larger Klenow fragment. The third is the 5′-3′ exonuclease activity that is located on the smaller fragment. It has been determined that each requires at least one divalent cation for activity.[46,47] The 3′-5′ exonuclease site serves a proofreading function for polymerization. When working concurrently, the error in polymerization can be reduced by tenfold.[48] The 3'-5' exonuclease site on DNA polymerase I acts to excise mismatched base pairs after polymerization.

Crystallographic data suggest that the 3'-5' exonuclease site requires at least two divalent cations for catalytic activity. Kinetic and mutagenesis studies on the Klenow fragment suggest one tight-binding site (using Mn^{2+}) with a $K_d \sim 2.5$ μM. When substrate analogs such as TMP or dGTP are added, two tight-binding sites were measured with a total $K_d \sim 8$ μM (for Mn^{2+}).[51] These two cations, labeled site A for the strong binding site and site B for the weaker binding site, are shown in Figure 6.5.[49,50] The $K_{\frac{1}{2}}(Mg^{2+})$ (the concentration of metal cation 6required for half-saturation) is 0.34 mM, which includes both A and B sites on the Klenow fragment. Note the greatly increased affinity for Mn^{2+}; $K_{\frac{1}{2}}(Mn^{2+}) \sim 4.2$ μM.[52]

There has been concern over the identity of such metal ions since both the kinetic studies and the crystal structure work used Mn^{2+} as the cofactor.[50,51] The enzyme is known to require both Zn^{2+} and Mg^{2+} for maximal activity.[47,53] Based on competition studies involving Co^{2+}, Zn^{2+}, Mn^{2+}, and Mg^{2+}, it is believed that the Zn^{2+} ion binds at site A while Mg^{2+} binds at site B.[50,52] From these structural studies it has been proposed that the reaction proceeds by way of a two-metal-ion catalytic scheme (Fig. 6.5).[54,55]

Mutagenesis experiments have been successfully employed to investigate the functional roles of the metal cofactors during turnover. Structural data show that the divalent cation at site A is bound by residues Glu-357 and Asp-355. After mutation to Ala, the exonucleolytic activity of the mutant was found to decrease by $\sim 10^5$-fold.[51] The Asp-424 to Ala mutation has been found to block metal binding to the B site, and also to weaken metal binding to site A by an order of magnitude.[50] Crystallographic analyses suggest that site B participates in catalysis, activating a nucleophilic water for hydrolysis of the phosphate, while metal A is thought to facilitate cleavage of the 3'-O phosphate bond by stabilization of the phosphate leaving group.[50,56] Some of these issues remain speculative, but nevertheless have proved useful in the design of test experiments.

6.4.3 Exonuclease III

The most common apurinic/apyrimidinc (AP) endonuclease in *E. coli* is exonuclease III. In fact, over 80% of all the AP endonuclease activity in the cell is accounted for by this enzyme.[58] The enzyme also displays ribonuclease H, exonuclease, and dephosphatase activities. Despite these multiple functions, the enzyme is a rather small monomeric protein with a molecular weight of 28 kDa.[57] Substrate recognition for exonucleolytic action appears to involve only those lesions where a secondary amine glycosidic linkage occurs; that is, tertiary linkages are not recognized as substrates.[58] An additional requirement for substrate recognition is the absence of base pairing at the reaction site.[58]

At this point it is uncertain how many active sites are present on exonuclease III, although it is thought that only one site is responsible for the protein's hydrolytic functions. Most attention has been focused on the AP activity, where the enzyme binds to an apurinic or apyrimidinic site, and after a ring-opening reaction (cata-

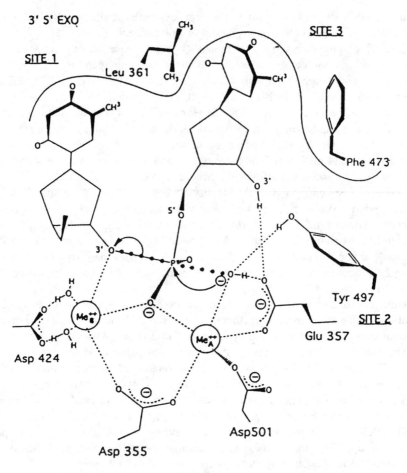

METAL SITE

Figure 6.5. The proposed magnesium-binding site in the active site of the 3′-5′ exonuclease site of *E. coli* DNA polymerase. The structure was solved with Mn^{2+} as metal cofactor; however, it is believed that Zn^{2+} and Mg^{2+} are the in vivo cofactors at sites *A* and *B*, respectively. Note that the magnesium and zinc ions are bridged by both Asp-355 and one phosphate oxygen. Catalysis is thought to be facilitated by magnesium polarizing the phosphate oxygen bond, making the phosphate susceptible to hydroxide attack.[54-56] Adapted from reference 55.

lyzed by the enzyme) the 1′-2⁰ amine shifts from a β conformation to a coplanar arrangement (with respect to the deoxyribose fragment) and forms an imine (Fig. 6.6). The presence of the enzyme is thought to stabilize this bonding arrangement and ease the steric congestion that can inhibit hydrolysis.[58,59] The functional role of magnesium has unfortunately not been very well characterized. One proposal is based on a discontinuity in an Arrhenius temperature plot. This suggests that upon

(a)

(b)

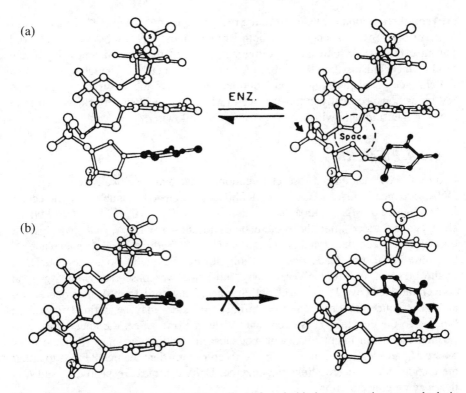

Figure 6.6 Schematic illustration of the effect of steric hinderance on the exonucleolytic action of exonuclease III at the 3′-terminus of a nucleotide strand. (a) Exonuclease III binds to the 3′-terminus and catalyzes the ring-opening reaction via an enzyme-stabilized iminium intermediate. Since the base lesion occurs at a terminal site, there is sufficient rotational freedom to facilitate bond cleavage. In the case of (b), exo III is unable to bind to the site and catalyze ring opening, since this requires planarity of the deoxyribose ring, which is hindered by the close proximity of the 5′ and 3′ nucleotides. Adapted from reference 58.

metal binding, two temperature-dependent conformations are available with a transition temperature ca. 25°C. At temperatures below 25°C a conformer exists that binds tightly to ds-DNA and is highly processive,[60] while at higher temperatures the enzyme changes to a distributive mode with bond scission as the rate-limiting step rather than association or dissociation.[58]

6.4.4 Eco RI and Eco RV

Both of these are commonly used sequence-specific magnesium-dependent restriction nucleases that show many functional similarities. Although the details of cleavage and recognition are not completely understood, much information concerning the thermodynamics and reaction specificity are known. Progress has been greatly assisted by the availability of a high-resolution crystal structure for Eco RI.[68] Both are type-II restriction enzymes since the recognition sequence is of the order of four to eight base pairs in length, and both strands of the DNA substrate are cleaved during catalysis. A fundamental difference between the two enzymes lies in the binding mechanisms. Eco RI has been found to bind to DNA with high specificity in the absence of magnesium cations, but does not cleave the substrate,[61] while Eco RV binds nonspecifically in the absence of Mg^{2+} and again cannot cleave substrate DNA.[62] These points will be addressed in the following.

6.4.4.1 Eco RI

Eco RI is a 31 kDa type-II restriction endonuclease that specifically recognizes the DNA sequence 5'-GAATTC-3'. Eco RI binds to cognate hexanucleotide molecules in the absence of Mg^{2+} with a dissociation constant $K_d \sim 10^{-11}$ M^{-1}.[63-65] Magnesium can be added after the protein binds to substrate DNA, and hydrolysis is observed between the G and A bases (leaving a 3'-hydroxyl and 5'-phosphate).[67]

Halford et al. noted a decrease in affinity for Eco RI binding to noncognate DNA in the presence of Mg^{2+}.[64] Previously, inhibition was observed using $MgCl_2$ at concentrations above 15 mM with activity increasing until 15 mM, then decreasing at higher concentrations.[67] Several affinity constants may be defined for cation binding to the enzyme. These constants, listed below, reflect the uncertainty concerning metal ion interactions with nucleic acid binding enzymes (Scheme 6.2), where M_E and M_D represent the Mg^{2+} concentration when the metal binds to either the enzyme alone or crosslinks the enzyme–DNA complex, respectively, and E is the enzyme concentration.[64]

$$M_E + E + M_D \overset{K_1}{\rightleftharpoons} E{\cdot}M_D \overset{K_2}{\rightleftharpoons} E_2{\cdot}M_D$$
$$\Updownarrow \; K_3 \qquad\qquad +E$$
$$E{\cdot}M_E$$

Scheme 6.2

The constant K_1 defines the binding of one magnesium ion to an enzyme molecule that is also bound to one DNA site, K_2 defines the binding of a second enzyme molecule, and K_3 defines cation binding to the enzyme where the complex cannot bind to DNA (Scheme 6.2). Note that: K_1 and K_2 were determined directly by analysis of experimental data, while K_3 was derived from K_1 and K_2.[64]

$K_1 = [E] [M_D]/[E.M_D] = 6$ nM

$K_2 = [E] [E.M_D]/[E_2.M_D] = 1.5$ nM

$K_3 = [E] [M_E]/[E.M_E] = 7 \pm 3$ nM

Several conclusions were made from these binding constants and related experiments. First, Mg^{2+} is involved to some degree in catalyzing the hydrolysis of the

DNA backbone. Second, Mg^{2+} binding results in a destabilization of the enzyme–substrate complex (a decrease in affinity of Eco RI for DNA). This destabilization is such that the free energy change for formation of the enzyme–DNA substrate complex is larger than for the formation of the enzyme–metal–substrate complex. That is, the decrease in activation energy is believed to occur when the cation binds, resulting in a lower affinity for the substrate. It has been proposed that Mg^{2+} destabilizes the enzyme–substrate complex, and liberates free energy that is channeled into catalytic activity (Fig. 6.7).

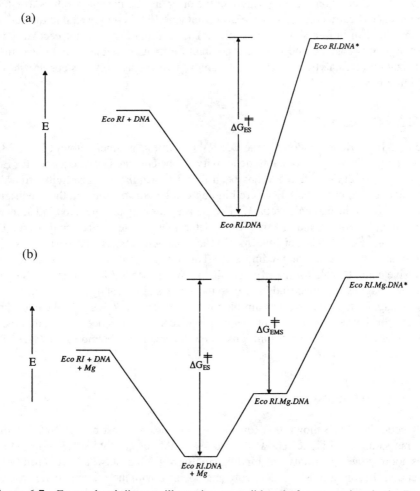

Figure 6.7   Energy-level diagram illustrating a possible role for magnesium in the Eco RI reaction profile. (a) Enzyme and substrate ($E + S$) stabilized in an ES complex. ES^* is much higher in energy and represents the reaction barrier to $E + P$ ($\Delta G_{ES}^{\neq}$). (b). The same scheme except magnesium ion is present and binds to the ES complex only. This *destabilizes ES* to a higher-energy form, lowering the effective transition-state energy to $\Delta G_{EMS}^{\neq}$, thereby facilitating product formation.

The structure of substrate-bound Eco RI has been crystallographically determined in the absence of Mg^{2+}.[68] Examination of the binding site provided evidence
for extensive interactions between one side of the DNA and amino acid side chains
(mediating 12 hydrogen bond contacts to the helix), while the opposite side of the
DNA duplex resides in a solvent channel. The scissile phosphodiester bond is
positioned at the end of this channel. Soaking the crystal in Mg^{2+}(aq) led to
activation of the enzyme and strand scission with no subsequent degradation of the
crystal. From these data it was concluded that the magnesium cation was able to
move down the solvent channel, independent of the bound DNA and bind at the
active site, whereupon the magnesium cofactor orients active site residues to estab-
lish the catalytically competent state of the enzyme.[68] It is proposed that the activation mechanism in free solution is similar. The fact that Mg^{2+} is not necessary for
DNA binding (cognate and noncognate sites) limits its possible roles to structural
organization or catalytic hydrolysis. A summary of the hydrolytic steps are shown
in Figure 6.8.

6.4.4.2 Eco RV

The type-II restriction endonuclease Eco RV is active as a dimer (subunit $M_r \sim 28.6$
kDa) and requires Mg^{2+} for catalytic activity.[69] The cognate DNA sequence for Eco
RV is 5'-GATATC-3'. Eco RV has been found to acquire its specificity from the
binding event, unlike Eco RI, where binding can be nonspecific and the specificity
is derived from the catalytic step.[70,71] Eco RI is apparently designed to bind at many
sites, but only with the aid of Mg^{2+} can it determine the site to effect hydrolysis. By
comparison, Eco RV is not able to bind at noncognate sites and derives its site
specificity from the initial binding step. This evidence alone suggests an almost
exclusive catalytic role for the Mg^{2+} cofactor. When Eco RV identifies the recognition site, magnesium must find its way to the active site for catalysis to occur. Based
on the magnitude of the magnesium binding constant for Eco RV (K_c; the equilibrium step for the metal ion binding to a noncognate site), it has been found that there
is a higher affinity for magnesium ions when the enzyme is bound to the GATATC
sequence.[71]

$$E{\cdot}S + M \overset{K_c}{\rightleftharpoons} E{\cdot}S{\cdot}M$$

Kinetic data have shown that for noncognate sites K_c must be $\geqslant 3$ mM,[71] while
enzyme saturation kinetics show K_c to be $\leqslant 1$ mM for cognate sites. Since Mg^{2+}
does not regulate specificity, the binding site for Mg^{2+} can show a lower affinity for
the cation. In solution, Mg^{2+} is largely unassociated with the enzyme in the absence
of cognate DNA binding sites. However, when a cognate site does exist, the enzyme
binds substrate with nanomolar affinity and reorganizes its active site in such a way
that increases Mg^{2+} binding on the order of 10^2-10^3-fold.[71]

Steady-state kinetics data illustrate the affinity of the enzyme for substrate (Table
6.2). Metal inhibition was not observed, although the concentration of Mg^{2+} was

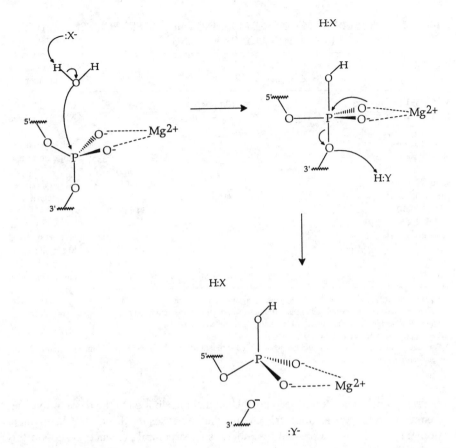

Figure 6.8 Putative general scheme for catalysis by both Eco RI and Eco RV. Both are thought to effect catalysis in the same manner. Differences in activity lie in the binding step, where Eco RI does not require Mg^{2+} to bind to its cognate site while Eco RV does require the cation. The scheme shows a currently unidentified acid (Y) and unidentified base (X) that are required to initiate catalysis. Magnesium is shown interacting with the phosphate oxygens, either polarizing the bonds or acting to stabilize intermediates. It is not known whether magnesium acts through its inner or outer coordination sphere.[61]

not taken above 10 mM. Kinetic studies have also been carried out using Mn^{2+} as cofactor. The main difference between Mg^{2+} and Mn^{2+} is that while Eco RV does not cleave at noncognate sites with Mg^{2+} as cofactor, Mn^{2+} relaxes this constraint, and the ability of the enzyme to discriminate cognate from noncognate sites is lowered by a factor of 10^4 (Table 6.3). This supports a model where the active site cannot bind magnesium until after substrate is bound. For Mn^{2+}, the enzyme binds DNA readily but with complete loss of specificity, indicating that the enzyme recognizes both cognate and noncognate sites. Figure 6.8 illustrates a general reaction mechanism for both Eco RI and Eco RV.

   The structural mechanisms used by Eco RV to increase its affinity for magnesium

Table 6.2 SUMMARY OF MAGNESIUM BINDING PARAMETERS FOR SELECTED
ENZYMES IN NUCLEIC ACID PROCESSING

Enzyme	Type of Site	$K_d{}^a$	$K_{1/2}{}^b$	Coordination number
DNA Pol I polymerase		100 μM		6[b]
RNA Pol I	Tight	1.9 μM[c]		Not crystallized
	Weak	200 μM		
Topoisomerase				Not crystallized
DNA gyrase	Catalytic		2 mM	Not determined
	ATPase		0.75 mM[d]	
Helicase			0.8 mM	Not crystallized
Ribonuclease H		60 μM		Not determined
DNA Pol I 3'-5' Exo			0.34 mM	6
Exonuclease III				Not crystallized
Eco RI		7 mM		Mg²⁺ not co-crystallized
Eco RV		≪1 mM		Not crystallized

[a] K_d is the dissociation constant for the metal binding site on the enzyme.
[b] $K_{1/2}$ is the magnesium concentration that affords half-optimal enzyme activity during the turnover of substrate.
[c] Mn^{2+} was used to determine the K_d for both the tight and weak sites.
[d] Determined from Figure 6.2.
[e] For cognate sites reaction rates could not be measured at higher Mg^{2+} concentrations since the rates show an ionic strength dependence.[71]

are currently not clear. It has been postulated that once bound to the recognition site, the enzyme undergoes an allosteric change that increases its affinity for the divalent cation[71]; however, at the present time there is no structural evidence to support this hypothesis.

6.5 Concluding Remarks

In this chapter we have examined the functional roles of Mg^{2+} as an activator of enzyme activity in nucleic acid biochemistry. Tables 6.2 and 6.4 summarize thermodynamic data and functional roles for essential divalent magnesium cofactors in

Table 6.3 MICHAELIS–MENTEN PARAMETERS FOR ECO RV. NOTE THE DECREASED REACTION RATE AND LOWER AFFINITY WHEN Mn²⁺ IS USED AS THE COFACTOR.[71]

Kinetic Constant	Mn^{2+}	Mg^{2+}
k_{cat} (min⁻¹)	0.04 ± 0.003	0.9 ± 0.05
K_m (nM)	1.7 ± 0.2	0.5 ± 0.1

Table 6.4 PROPOSED FUNCTIONS FOR MAGNESIUM IONS
IN NUCLEIC ACID PROCESSING ENZYMES

| Enzyme | Allosteric (enzyme)[a] | Allosteric (substrate)[b] | ATPase | Phosphate Hydrolysis[c] | |
				Stabilize rn. Intermediates	Lewis Acid Catalaysis
DNA Pol I (polym)		X		X	
DNA Pol I (3'-5' exonuclese)				X	
RNA Pol I	X		X		
Topoisomerase I[d]	X	X		X	
DNA Gyrase			X	X[e]	X[e]
Helicase		X	X		
Ribonuclease H				X[e]	X[e]
Exonuclease III	X				
Eco RI	X[f]				X
Eco RV					X

[a] In these cases Mg^{2+} serves as an allosteric regulator of enzyme activity, possibly organizing active site structure.
[b] Mg^{2+} acts to modify the substrate to facilitate either the catalytic reaction or enzyme binding.
[c] Indicates a direct role in catalysis. Two proposed mechanisms involve either the stabilization of a transition state/ intermediate, or Lewis acid catalysis.
[d] It is not known whether Mg^{2+} changes enzyme or substrate conformation.
[e] An "X" in each spot indicates a lack of agreement on cation function.
[f] Mg^{2+} organizes the active-site residues after substrate binding to the protein.

those enzymes that we have discussed earlier. Close inspection of published data demonstrates only a few fundamental roles that relate to structure stabilization and catalysis (either by direct coordination, or electrostatic stabilization of intermediates). The numbers of active metal cofactors are typically only one or two. Additional very weakly bound ions are unlikely to play substantive roles, most likely reflecting adventitious binding sites. In both Sections 6.2 and 6.3, the metal cofactors were observed to play significant structural roles, often stabilizing important conformations or structural intermediates. In Section 6.4 the hydrolytic roles of the metal cofactor were emphasized; however, the comparison of Eco RI and Eco RV site recognition suggests that metal cofactors play important but subtle roles in the regulation of protein–nucleic acid contacts. This should be a fruitful area for further study.

References

1. Suh, W. C.; Leirmo, S.; Record, M. T., *Biochemistry* **1992**, *31*, 7815–25.

2. Focher, F.; Verri, A.; Maga, G.; Spadari, S.; Hübscher, U., *F.E.B.S.* **1990**, *259*, 349–52.

3. Mullen,, G. P.; Serpersu, E. H.; Ferrin, L. J.; Loeb, L. A.; Mildvan, A. S., *J. Biol. Chem.* **1990**, *265*, 14327–34.

4. Polesky, A. H.; Steitz, T. A.; Grindley, N.D.F.; Joyce, C. M., *J. Biol. Chem.* **1990**, *265*, 14579–91.

5. Ferrin, L. J.; Mildvan, A. S., *Biochemistry* **1985**, *24*, 6904–13.

6. Ollis, D. L.; Brick, P.; Hamlin, R.; Xuong, N. G.; Steitz, T. A., *Nature* **1985**, *313*, 762–66.

7. Sigel, H., *Eur. J. Biochem.* **1987**, *165*, 65–72.

8. Koren, R.; Mildvan, A. S., *Biochemistry* **1977**, *16*, 241–49.

9. Singer, P. T.,; Wu, C.-W., *J. Biol. Chem.* **1988**, *263*, 4208–14.

10. Roe, J.-H.; Record, Jr., M. T., *Biochemistry* **1985**, *24*, 4721–26.

11. Suh, W. C.; Leirmo, S.; Record, Jr., M. T., *Biochemistry* **1992**, *31*, 7815–25.

12. Wang, J. C., *Ann. Rev. Biochem.* **1985**, *59*, 665–97.

13. Osheroff, N., *Pharmacol. Ther.* **1989**, *41*, 223–41.

14. Tse, Y-C.; Kirkegaard, K.; Wang, J. C., *J. Biol. Chem.* **1980**, *255*, 5560–65.

15. Der Garabedian, P. A.; Mirambeau, G.; Vermeersch, J. J., *Biochemistry* **1991**, *30*, 9940–47.

16. Domanico, P. L.; Tse-Dinh, Y. C., *J. Inorg. Biochem.* **1991**, *42*, 87–96.

17. Adrian, M.; ten Heggler-Bordier, B.; Walhi, W.; Stasiak, A.; DuBochet, J., *E.M.B.O.J.* **1990**, *9*, 4551–54.

18. Kim, S.; Cowan, J. A., *Inorg. Chem.* **1992**, *31*, 3495–96.

19. Rau, D. C.; Gellert, M.; Thoma, F.; Maxwell, A., *J. Mol. Biol.* **1987**, *193*, 555–69.

20. Osheroff, N.; Shelton, E. R.; Brutlag, D. L., *J. Biol. Chem.* **1983**, *258*, 9536–43.

21. Wang, J. C., *Ann. Rev. Biochem.* **1985**, *54*, 665–97.

22. Wigley, D. B.; Davies, G. J.; Dodson, E. J.; Maxwell, A.; Dodson, G., *Nature (London)* **1991**, *351*, 624–29.

23. Reece, R. J.; Dauter, Z.; Wilson, K. S.; Maxwell, A.; Wigley, D. B., *J. Mol. Biol.* **1990**, *215*, 493–95.

24. Lebeau, L.; Regnier, E.; Schultz, P.; Wang, J. C.; Mioskowski, C.; Oudet, P., *F.E.S.B. Lett.*, **1990**, *267*, 38–42.

25. Osheroff, N., *Biochemistry* **1987**, *26*, 6402–6.

26. Black, C. B.; Huang, H.-W.; Cowan, J. A., *Coord. Chem. Rev.*, **1994**, in press.

27. Higgins, N. P.; Cozzarelli, N. R., *Nucleic Acids Research* **1982**, *10*, 6833–47.

28. Abdel-Monem, M.; Durwald, H.; Hoffman-Berling, H., *Eur. J. Biochem.* **1976**, *65*, 441–49.

29. Matson, S. W.; Kaiser-Rogers, K. A., *Ann. Rev. Biochem.* **1981**, *50*, 233–60.

30. Steinmetz, E. J.; Brennan, C. A.; Platt, T., *J. Biol. Chem.* **1990**, *265*, 18408–13.

31. Ray, B. K.; Lawson, T. G.; Kramer, J. C.; Cladras, M. H.; Grifo, J. A.; Abramson, R. D.; Merrick, W. C.; Thach, R. E.; *J. Biol. Chem.* **1985**, *260*, 7651–58.

32. Company, M.; Arenas, J.; Abelson, J., *Nature (London)* **1991**, *349*, 487–93.

33. Brennan, C. A.; Steinmetz, E. J., Spear, P.; Platt, T., *J. Biol. Chem.* **1990**, *265*, 5440–47.

34. Chao, K.; Lohman, T., *J. Mol. Biol.* **1991**, *221*, 1165–81.

35. Poll, E.H.A.; Benbow, R. M., *Biochemistry* **1988**, *27*, 8701–6.

36. Katayanagi, K.; Miyagawa, M.; Matsushima, M.; Ishikawa, M.; Kanaya, S.; Ikehara, M.; Matsuxaki, T.; Morikawa, K., *Nature (London)* **1990**, *347*, 306–9.

37. Davies II, J. F.; Hostomska, A.; Hostomsky, Z.; Jordan, S. R.; Matthews, D. A., *Science* **1991**, *252*, 88–95.

38. Nakamura, H.; Oda, Y.; Iwai, S.; Inoue, H.; Ohtsuka, E.; Kanaya, S.; Kimura, S.; Katsuda, C.; Katayanagi, K.; Morikawa, K.; Miyashiro, H.; Ikehara, M., *Proc. Natl. Acad. Sci. U.S.A.*, **1991**, *88*, 11535–39.

39. Kanaya, S.; Katsuda-Nakai, C.; Ikehara, M., *J. Biol. Chem.* **1991**, *266*, 11621–27.

40. Kanaya, S.; Kimura, S.; Miura, Y.; Sekiguchi, A.; Iwai, S.; Inoue, H.; Ohtsuka, E.; Ikehara, M., *J. Biol. Chem.* **1990**, *265*, 4615–21.

41. Yang, W.; Hendrickson, W. A.; Crouch, R. J.; Satow, Y., *Science* **1990**, *249*, 1398–405.

42. Huang, H.-W.; Cowan,, J. A., *Eur. J. Biochem.* **1994**, *219*, 253–260.

43. Jou, R.; Cowan, J. A., *J. Am. Chem. Soc.* **1991**, *113*, 6685–86.

44. Black, C. B.; Cowan, J. A., manuscript submitted.

45. Black, C. B.; Cowan, J. A., *J. Am. Chem. Soc.* **1994**, *116*, 1174–78.

46. Bessman, M. J.; Lehman, I. R.; Simms, E. S.; Kornberg, A., *J. Biol. Chem.* **1958**, *233*, 171–77.

47. Lehman, I. R.; Richardson, C. C., *J. Biol. Chem.* **1964**, *239*, 233–41.

48. Loeb, L. A.; Kunkel, T. A., *Ann. Rev. Biochem.* **1982**, *52*, 429–37.

49. Ollis, D. L.; Brick, P.; Hamlin, R.; Xuong, N. G.; Steitz, T. A., *Nature (London)* **1985**, *313*, 762–66.

50. Derbyshire, V.; Freemont, P. S.; Sanderson, M. R.; Beese, L.; Friedman, J. M.; Joyce, C. M.; Steitz, T. A., *Science* **1988**, *240*, 199–201.

51. Mullen, G. P.; Serpersu, E. H.; Ferrin, L. J.; Loeb, L. A.; Mildvan, A. S., *J. Biol. Chem.* **1990**, *265*, 14327–34.

52. Han, H.; Rifkind, J. M.; Mildvan, A. S., *Biochemistry* **1991**, *30*, 11104–8.

53. Kornberg, A. *DNA Replication* 1980 W. H. Freeman, San Fran., CA.

54. Steitz, T. A.; Steitz, J. A., *Proc. Natl. Acad. Soc. U.S.A.*, **1993**, *90*, 6498–502.

55. Beese, L. S.; Steitz, T. A., *E.M.B.O.J.* **1991**, *10*, 25–33.

56. Freemont, P. S.; Friedman, J. M.; Beese, L. S.; Sanderson, M. R.; Steitz, T. A., *Proc. Natl. Acad. Sci. U.S.A.* **1988**, *85*, 8924–28.

57. Weiss, B., *J. Biol. Chem.* **1976**, *251*, 1896–901.

58. Kow, Y. W., *Biochemistry* **1989**, *28*, 3280–87.

59. Weiss, B., *Enzymes* **14** (3rd edition), 1981, 203–31.

60. Wu, R.; Ruben, G.; Siegal, B.; Jay, E.; Spielman, P.; Tu, C. D., *Biochemistry* **1976**, *15*, 734–40.

61. Jeltsch, A.; Jurgen, A.; Gunter, M.; Pingoud, A., *F.E.B.S.* **1992**, *304*, 4–8.

62. Taylor, J. D.; Badcoe, I. G.; Clarke, A. R.; Halford, S. E., *Biochemistry* **1991**, *30*, 8743–53.

63. Modrich, P., *Q. Rev. Biophys.* **1979**, *12*, 315.

64. Halford, S. E.; Johnson, N. P., *Biochem. J.* **1980**, *191*, 593–604.

65. Rosenberg, J. M.; Boyer, H. W.; Greene, P. J., *Gene Amplification and Analysis: Volume I*,

Restriction Endonucleases, Chirikjian, J. G. (Ed.), Elsevier/North-Holland, Amsterdam, 1981, p. 131.

66. Hedgpeth, J.; Goodman, H. M.; Boyer, H. W., *Proc. Natl. Acad. Sci. U.S.A.* **1972,** *69,* 3448–52.

67. Roulland-Dussoiz, D.; Yoshimori, R.; Greene, P.; Betlach, M.; Goodman, H. M.; Boyer, H. W., *American Society for Microbiology, Conference on Bacterial Plasmids,* Schlessinger, D. (Ed.), ASM Press, Washington, 1974, pp. 187–98.

68. McClarin, J. A.; Frederick, C. A.; Wang, B. C.; Greene, P.; Boyer, H. W.; Grable, J.; Rosenberg, J. M., *Science* **1986,** *234,* 1526–42.

69. Luke, P. A.; McCallum, S. A.; Halford, S. E., *Gene Amplif. Anal.* **1987,** *5,* 183–205.

70. Newman, P. C.; Williams, M. D.; Cosstick, R.; Seela, F.; Connolly, B. A., *Biochemistry* **1990,** *29,* 9902–10.

71. Taylor, J. D.; Halford, S. E., *Biochemistry* **1989,** *28,* 6198–207.

CHAPTER 7

Magnesium-Dependent Enzymes in General Metabolism

C. B. Black and J. A. Cowan

7.1 Introduction

Magnesium is an essential element that is utilized by living cells in a variety of biochemical and biophysical roles to maintain life processes. In this chapter we will examine the important structural and catalytic function of divalent magnesium in regulating the activity of important enzymes in major metabolic pathways; particularly of those enzymes involved in glycolysis, amino acid synthesis, photosynthesis, the citric acid cycle, and reactions of general importance such as phosphoryl transfer. Magnesium activation of enzymes in pathways relating to nucleic acid biochemistry is discussed in Chapter 6. To emphasize the overall importance of magnesium in cellular biochemistry, this chapter is subdivided into sections relating to specific reaction pathways. We will review a large body of kinetic and crystallographic data and attempt to derive mechanistic insight on the chemistry of the magnesium cofactor in a variety of metalloenzymes.

7.2 Glycolytic Enzymes

The importance of divalent magnesium in cellular metabolism is nowhere more obvious than in the glycolytic cycle (Fig. 7.1). Most of the key enzymes in this pathway are magnesium dependent. Our attention will focus on those cases where the chemistry of the metal cofactor has been addressed by kinetic and/or structural studies.

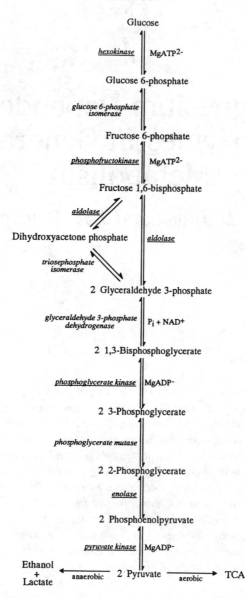

Figure 7.1   Glycolytic cycle illustrating the general importance of the magnesium cofactor. Magnesium-dependent enzymes are underlined.

7.2.1 Xylose Isomerase

Xylose isomerase is not a bona fide glycolytic enzyme as defined by the cycle in Figure 7.1; however, it can process glucose by converting it to the five-membered ring isomer, ketose. The enzyme is tetrameric, containing four identical 43 kDa subunits that define the two binding sites for magnesium ions (separated by less than 5 Å) that participate in the catalytic isomerization of aldose D-xylose to ketose D-xylose (Fig. 7.2).[1] Xylose isomerase also converts D-glucose to D-fructose, although the efficiency is lower.[1]

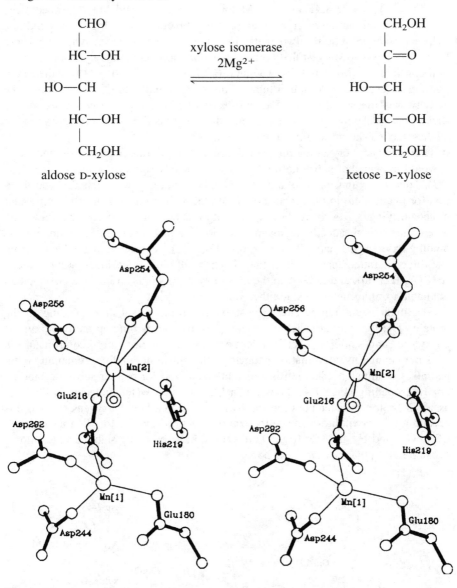

Figure 7.2 Stereoview of the coordination geometry of metal sites (1) and (2) for the Mn derivative of *Arthrobacter* xylose isomerase. Note the bridging Glu-216 between the two metal ions. This structure was solved without substrate. The Mg^{2+}-substituted enzyme showed the same contacts including the unusual coordination environment around metal site (1).[1] Adapted from ref. 9.

Kinetic parameters for metal-promoted enzyme activity (Mn^{2+}, Co^{2+}, and Mg^{2+}) have been determined by several groups.[1-4] All show agreement in the magnitude of the relative binding strengths of the three cations. For the isomerase isolated from *E. coli,* the concentration of metal ion that yields half-maximal velocity ($K_\frac{1}{2}$) is 4.8, 8.4, and 80 μM for Mn^{2+}, Co^{2+}, and Mg^{2+}, respectively. Through a combination of crystallographic and kinetic data, it has been found that xylose isomerase requires two metal ions for full functional activity.[1,3,4] These kinetic studies also suggest that both cations bind non-cooperatively, with dissociation constants $K_1(Mg^{2+}) \leq 10^{-6}$ M and $K_2(Mg^{2+}) = 6 \times 10^{-5}$ M (for the *Streptomyces* sp enzyme).[4] Although tighter binding is found with Co^{2+} and Mn^{2+}, the activity was measured to be 45% and 8% of the wild type, respectively, for the *Arthrobacter* enzyme.[3] Inhibition kinetics have not been observed by any group in studies carried out at high metal concentration.

On the basis of crystallographic evidence and supporting mutagenesis studies,[1,5] it would appear that substrate binding occurs at only one of the metal sites (Fig. 7.3). This site is unusual insofar as the crystal structure shows tetrahedral coordination for magnesium in the absence of substrate, with the nearest-neighboring water molecule lying 3.2 Å away. Four-coordinate magnesium is extremely rare,[6,7] and when substrate is added the magnesium ion adopts octahedral coordination by binding oxygens O1 and O2 from xylose (Fig. 7.3).[1] Corroborating evidence from site-directed mutagenesis shows that deletion of binding residues around site 1 (E181A, E217D), and D292A in the *E. coli* system) yields an inactive enzyme that is unable to bind metal ions at either site.

The magnesium ion at site 1 most likely promotes substrate binding to the protein. The ion may also help to orient the substrate by binding the two oxygens on xylose, and stabilize reaction intermediates. In contrast, the second metal ion does not change its coordination geometry after substrate binding, remaining octahedrally coordinated. This ion is apparently required for neither substrate binding nor ring opening (Fig. 7.4); however, a Mg^{2+}-bound water molecule is likely to assist in proton transfer between the hydroxyl and carbonyl oxygens.[8] Since this second ion is essential for the isomerization reaction (that is, for H^+ transfer from C2 to C1, and H^+ transfer from O2 to O1), it most probably stabilizes the anionic substrate and polarizes the C–O bonds (Fig. 7.4).

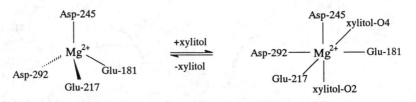

Figure 7.3 Metal site 1 of xylose isomerase from *Actinoplanes missouriensis* before and after binding of the substrate analog (xylitol). Glu-217 bridges the two magnesium ions that form the active site. Note the unusual coordination environment when substrate is absent (no water molecules were seen within 4 Å.[1]

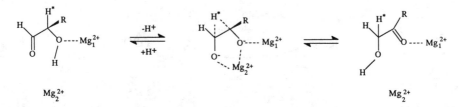

Figure 7.4 Proposed mechanism for the metal-catalyzed 1,2-hydride shift in *Arthrobacter* xylose isomerase. The labile hydrogen is starred. Other ligands binding to the Mg^{2+} ions are not shown. Note that Mg_2^{2+} is thought to be the catalytic ion, while the other most likely plays a structural role.[9]

Kinetic data for Mg^{2+}-activated xylose isomerase suggest that the first step in catalysis is the binding of Mg^{2+} to site 1 (having the higher binding affinity), which facilitates substrate binding. Subsequently, the second cation binds and turnover is initiated.[3] The rate-limiting step is most likely isomerization of xylose. Figure 7.4 summarizes the major steps in metal-ion activation of xylose isomerase.

7.2.2 Enolase

Enolase is a homodimeric protein that catalyzes the interconversion of 2-phosphoglycerate and phosphoenolpyruvate. Each subunit of the yeast dimer requires magnesium, both to maintain structural integrity and for catalysis.[11-13]

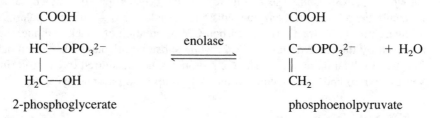

2-phosphoglycerate   phosphoenolpyruvate

Metal activation of enolase has been well characterized through crystallographic and NMR studies. Two divalent metal ions are required for full activation of the enzyme, with the tighter-binding site serving as the structural site. Kinetic studies suggest that this cation promotes substrate binding as a result of a structural rearrangement (Fig. 7.5),[13,14] which also favors binding of the second metal ion (or catalytic ion) at site 2. The role of this metal cofactor is not as well understood, since the cation is only weakly bound, but it is believed to regulate catalysis. One should note that the proposed metal-binding sites on the crystallographically characterized enzyme are tentative, since only one of the metal cofactors has been co-crystallized with the enzyme.[10,15] Binding constants for both sites are noted in Table 7.1.

On the basis of kinetic data there appears to be additional weak binding sites that inactivate the enzyme, resulting in inhibition kinetics for metal ion concentrations

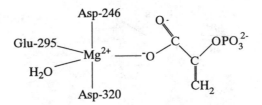

Figure 7.5 The crystallographically characterized magnesium-binding site in yeast enolase,[10] which is believed to serve a structural role. Note the direct coordination of the substrate and retention of an inner-sphere water molecule. PEP = phosphoenol pyruvate.

above 500 μM.[11] To rationalize this inhibition, the data were best fit to an equation that included an additional two metal binding sites (giving four sites in total). This can explain the gradual loss of activity, although the location and coordination of these other metal sites is currently unknown. Previously, an inhibitory site was found ($K_d \sim 2$ mM) from a kinetic study of rabbit muscle enolase.[16]

Enzyme activation requires two metal-binding steps. First, on the basis of crystallographic and NMR data,[a] it is believed that the C3-OH of 2-phosphoglycerate coordinates to the high-affinity ion at site 1 (Fig. 7.5).[17] This is accompanied by a structural change of the enzyme that promotes binding of a second catalytic metal ion, which then facilitates the isomerization reaction. The role of this site is poorly defined since it has not been structurally characterized. A mechanism similar to that shown for xylose isomerase in Figure 7.4 has been proposed, whereby a water molecule neighboring the enzyme substrate complex (2.6 Å away) is polarized,[10] allowing proton abstraction from carbon C2. Regulation of attack is thought to be achieved by the catalytic ion at site 2, while substrate orientation is carried out by the structural cation at site 1. The C2 proton can be transferred to Gln-168 (in the yeast enzyme) and subsequently to the OH group on 2-phosphoglycerate.

7.2.3 Pyruvate Kinase

Pyruvate kinase is a glycolytic enzyme that catalyzes the essentially irreversible conversion of phosphoenolpyruvate to pyruvate by transfer of phosphate to ADP. The two most common isozymes, cat muscle and rabbit muscle pyruvate kinase, both function as 240 kDa tetramers. Two divalent magnesium ions and a monovalent cation (K$^+$) bind to each subunit.

---

[a] Abbreviations used: NMR, nuclear magnetic resonance; ADP, adenosine diphosphate; ATP, adenosine triphosphate; DNA, deoxyribonucleic acid; PEP, phosphoenolpyruvate; GTP, guanosine triphosphate; TCA, tricarboxylic acid; RuBisco, ribulose-1,5-bisphosphate carboxylase; Da, Dalton; NAD$^+$, nicotinamide adenine dinucleotide; IDH, isocitrate dehydrogenase.

Table 7.1 K_m AND HALF-METAL-SATURATION
CONSTANTS FOR YEAST ENOLASE

Binding Parameters	Constants (mM)
K_{Mg}(structural)	<0.02
K_{Mg}(catalytic)	<0.05
$K_M{}^a$	0.14^b

[a] The Michaelis–Menten parameter.
[b] Adapted from Ref. 11.

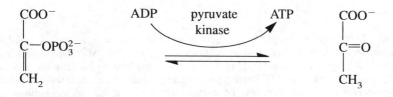

One of the divalent cations binds strictly to the enzyme and the phosphate of phosphoenolpyruvate, while the other chelates to ADP, but also binds to the enzyme at an allosteric site (Fig. 7.6).[18] Despite extensive study, the specific function of the bound cations remains unclear. For example, K^+ appears to regulate the stability and tertiary structure of the enzyme in an as yet unknown manner.[19] The enzyme is active only when two divalent cations and one monovalent cation are bound. Through single-angle neutron scattering data it has been shown that the active form is conformationally distinct from the inactive form, involving rotation of one domain relative to another. This closes a cleft between the domains that is thought to constitute the active site, where both metals and substrate bind.[20] From the crystal structure of cat muscle pyruvate kinase it would appear that phosphoenolpyruvate binds to the enzyme only through contacts with divalent cations, and not by direct interactions with protein side chains.[18] The ligands to the catalytic metal ion include Ala-292 (carbonyl), Glu-271, and Arg-293 (numbering in cat muscle). When the magnesium associated with the ADP substrate (MgADP) binds, the Lys-269 side chain moves closer to the phosphoenolpyruvate binding site, and away from the cation, where it can initiate catalysis.[18] The second magnesium, associated with ADP (or ATP) has minimal enzyme contacts (Asp-112). This cation is believed both to facilitate phosphate transfer and to effect structural changes through subunit reorientation.

Kinetic studies of pyruvate kinase have for the most part used Mn^{2+} and Cr^{3+}-adenosine nucleotides as metal cofactors,[21–23] rather than the spectroscopically silent magnesium ion. Results from one of the few kinetic studies utilizing divalent magnesium as cofactor (for the plastidic and cystolic pyruvate kinase from *Se-*

(a)   (b)

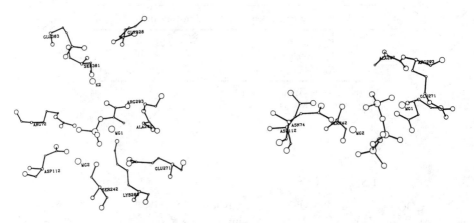

Figure 7.6   Active-site schematic of cat muscle pyruvate kinase. (a) The cocrystallized phosphoenolpyruvate substrate (shown in the center) neighbors the ADP-complexed cation $[Mg_2^{2+}]$ and the enzyme-bound cation $[Mg_1^{2+}]$. The enzyme-bound potassium ion is seen at the top of the diagram. (b) The ATP binding site with both divalent cations is shown. Binding interactions between ATP (center) and the enzyme are mediated through the cations, and Ser-242. Adapted from ref. 18.

lenastrum minutum) are noted in Table 7.2. Note, however, that the dissociation constant ($K_d \sim 0.75$ mM) obtained for Mn^{2+}-substituted pyruvate kinase is similar to that measured for Mg^{2+}.[21]

Based on the kinetic and crystallographic evidence currently available, the allosteric divalent cation (associated with ADP or ATP) appears to orient and polarize the phosphoryl group undergoing transfer, and effects a conformational change to an active state.[22] The functional role is less well understood in terms of the structural details, although one possibility for orienting and polarizing the γ-phosphate is summarized in Figure 7.7. Prior to reaction, binding of MgADP breaks a salt bridge

Table 7.2   INHIBITION AND HALF-SATURATION CONSTANTS FOR MgATP AND Mg^{2+}, RESPECTIVELY, FOR *SELENASTRUM MINUTUM* PLASTIDIC (P) AND CYSTOLIC (C) PYRUVATE KINASE[24]

	Pyruvate Kinase (p)	Pyruvate Kinase (c)
$K_{1/2}{}^a$	0.55 mM	0.85 mM
n^b	1.3	1.6
$I_{1/2}$ (Mg-ATP)^c	1.7 mM	6.8 mM

a   This is the metal concentration required for half-saturation.
b   Hill coefficient for Mg^{2+} cofactor.
c   This hs the MgATP concentration required for half-inhibition.

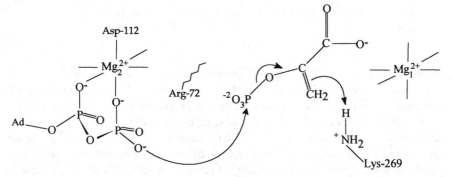

Figure 7.7 A proposed reaction mechanism for pyruvate kinase. It is thought that Mg_1^{2+} provides electrostatic stabilization during formation of the covalent intermediate. In addition, Mg_1^{2+} is thought to repel the active Lys residue toward the double bond to facilitate bond formation. Note that Mg_2^{2+} is complexed to ADP and aids in phosphate transfer.[18,19]

between Asp-112 and Arg-72 (in cat muscle), forming a Mg^{2+}–Asp bond and leaving the Arg residue free to interact with the γ-phosphate.[18,25] The catalytic cation appears to be responsible for both charge neutralization and the correct positioning of the lysine residue. As described, these results suggest that the metal cofactors serve to position specific residues in the active site to accept the substrate and effect catalysis, although this does not exclude the possibility of Lewis acid catalysis by direct coordination to phosphate.

7.3 Phosphorylation and Dephosphorylation

Phosphate and phosphoryl transfer reactions are ubiquitous in cellular biochemistry, and require magnesium ion as an essential cofactor.[6b] Distinctions between enzymes in this class can be related to the function of magnesium. Generally divalent magnesium will either make its principal bonding contacts with the triphosphate/ diphosphate molecule, or with side chains in the enzyme. In those cases where crystallographic data are available, the dominant binding mode can usually be deduced. Inasmuch as the catalytic role for Mg^{2+} in the simple case of ATP hydrolysis has been discussed in detail elsewhere,[6] this review chapter will focus on more general reactions, and will emphasize details pertaining to the role of enzyme–metal cofactor contacts. These points will be illustrated through several examples that are fairly well developed.

7.3.1 Alkaline Phosphatase

Dimeric alkaline phosphatase catalyzes the hydrolysis of 5′-monophosphates from DNA in addition to the transfer of inorganic phosphate to alcohols.[26] The metal ion requirement for catalysis consists of two zinc ions (the catalytic ions) and one magnesium ion (catalytic and structural).

$$E + ROPO_3^{2-} \rightleftharpoons E \cdot ROPO_3^{2-} \searrow E\text{-}PO_3^{2-} \cdots E + HPO_4^{2-}$$
$$\underset{R\dot{O}H}{\searrow} \quad H_2O$$

The crystal structure of the *E. coli* enzyme shows an octahedrally coordinated magnesium situated 4.9 Å from Zn_2 (Fig. 7.8) and 7.1 Å away from Zn_1. The ligand environment emphasizes the tendency of magnesium to retain at least part of the inner solvent shell, with three enzyme contacts (Thr-155, Glu-322, and Asp-51) and three water molecules (Fig. 7.8).[27] Crystallographic data obtained for the enzyme cocrystallized with phosphate ion show the phosphate bound at the active site, contacting the magnesium ion through a hydrogen bond.[27] The distance between phosphate oxygen and the metal-bound water is ~3 Å, making the total Mg^{2+}– phosphate distance on the order of 5 Å.[27] Although the magnesium ion does contact the substrate by hydrogen-bond formation through a metal-bound water, the role of magnesium in alkaline phosphatase is not as clear as that of the two zinc ions that form the catalytic core. It is accepted that divalent magnesium regulates the activity of the catalytic zinc site, since the activity increases by a factor of 1.2 when Mg^{2+} is present.[28] Also, it has been shown through proton–tritium exchange studies that Mg^{2+} lends conformational stability to the enzyme.[28,29] Based on these experiments a purely structural or conformational role has been assigned to magnesium; however, recent kinetic studies of site-directed mutants suggest that Mg^{2+} stabilizes the active conformation of the enzyme and has a direct effect upon catalysis.[30,31] Furthermore, pre-steady-state kinetics show a different rate-limiting step for the Mg^{2+}-free enzyme versus the Mg^{2+}-stimulated enzyme, suggesting that Mg^{2+} influences the step where the covalent bond between alkaline phosphatase and the phosphate is cleaved. Although a fast conformational change has not been excluded, the data support an intimate role for magnesium in catalysis.

7.3.2 Ha-Ras p21

The family of ras genes code for regulatory proteins that bind guanine nucleotides (G proteins). Ha-Ras p21 binds a divalent magnesium–GTP complex and has been

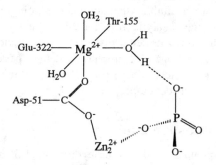

Figure 7.8 Magnesium binding site in *E. coli* alkaline phosphatase showing Mg^{2+} connectivity to the cocrystallized phosphate and the zinc ion at site 2. Note the retention of three inner-sphere water molecules on Mg^{2+} after coordination of substrate to the binuclear zinc center.[27,31]

the subject of extensive crystallographic characterization, although mechanistic details of its functional chemistry have yet to be carefully examined. The crystal structure, now refined to 1.35 Å, shows the bound complex of a substrate analog GppNp (where N is an imidazole spacer), with magnesium ion coordinated equivalently to each of the γ- and β-phosphate oxygens ($r_{Mg-O} \sim 2.2$ Å each). The remaining ligands to magnesium include two axial waters and hydroxyls from Thr-35 and Ser-17 (Fig. 7.9).[32,33] Time-resolved crystallographic data indicate that a change in metal-ion coordination triggers substantial changes in structure during hydrolysis.[33] Although the details are still unclear, it is now known that upon hydrolysis of the γ-phosphate, Mg^{2+} coordinates to the carbonyl group of Asp-57, leaving only one phosphate ligand (Fig. 7.9).[33] In addition to an added aspartate contact, the Thr-35 contact is lost. These differences in the coordination sphere around magnesium lead to large conformational changes in an "effector loop" of approximately seven residues in length. This loop is believed to be responsible for the binding of GAP (GTPase activating protein), which activates Ha-Ras p21 for catalysis.[33,34] From these structural data, and previous mechanistic evidence,[35] either a protein-bound water situated directly opposite the leaving group or the β-phosphate oxygen may be the attacking nucleophile. After loss of the nucleophilic water molecule, a five-coordinate phosphate intermediate is formed.[33] It is likely that magnesium increases the electrophilicity of the γ-phosphate center through coordination (Fig. 7.9) and stabilizes the product after transfer.[36]

7.3.3 Che Y

Che Y is a signal transduction protein in chemotactic bacteria that is involved in phosphoryl group transfer. In common with many such enzymes, magnesium is required for both the phosphoacceptor and phosphatase activities.[37] Aside from extensive crystallographic structural analyses, the chemistry of the magnesium cofactor has not yet been probed in great depth.

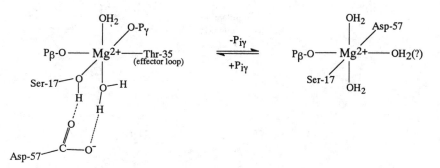

Figure 7.9   The magnesium binding site in Ha-Ras p21 before and after γ-phosphate transfer. Note the retention of two inner-sphere water molecules. Also, Asp-57 replaces the γ-phosphate after transfer. Thr-35, which is part of the effector loop, moves to ~4 Å distance from Mg^{2+} after transfer. The replacement of a metal-bound H_2O by threonine has not been confirmed by crystallography.[32,34]

Scatchard analyses and steady-state kinetic studies show magnesium to be catalytically important for phosphate transfer, not simply for nucleotide chelation. Fluorescence studies have provided a magnesium binding constant (K_d) of 500 ± 80
μM.[37] Moreover, the Scatchard analysis revealed $n = 1.05 \pm 0.09$ for magnesium,
which correlate with the one site identified in the crystal structure.[38,39] Mn^{2+},
Co^{2+}, and Zn^{2+} substitutions have been carried out and show dissociation constants
that are an order of magnitude higher than that determined for Mg^{2+} binding.
Inhibition kinetics were not observed, although the influence of high metal ion
concentrations (above 5 mM) have not been examined. In support of the kinetic
analysis, the structure (at 1.7 Å) shows a possible divalent cation binding site that is
surrounded by a group of acidic residues (Fig. 7.10).[38] The cation is thought to
neutralize the negative charge in the pocket and possibly stabilizes a pentacoordinate phosphoryl intermediate.[39] Specifically, it is proposed that magnesium anchors
an oxygen on Asp-57 to allow rotation of the other oxygen about the $C–O–Mg^{2+}$
axis. This rotation provides an in-line path for attack of the phosphate and formation
of a five-coordinate phosphoryl intermediate (Fig. 7.11).[39] Many of the mechanistic
details for the reaction path of Che Y–catalyzed phosphoryl transfer are based on
the available crystal structure and have been deduced from model building and
comparison with related systems.

7.4 Amino Acid Synthesis

7.4.1 Glutamine Synthetase

Glutamine synthetase is a dodecamer that catalyzes the formation of glutamine from
glutamate with accompanying hydrolysis of ATP.

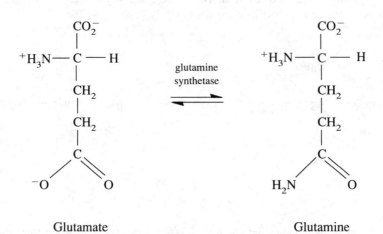

Glutamate   Glutamine

Crystallographic studies have identified two metal binding sites with an internuclear
distance of 5.8 Å.[40] Each lies at the interface of two domains, yielding a total of six

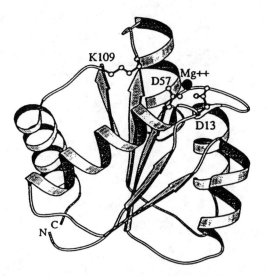

Figure 7.10   Schematic of Che-Y illustrating secondary structure and the Mg^{2+} binding site. Note the acidic residues that constitute the binding pocket and define its solvent accessibility. Adapted from ref. 39.

active sites. Kinetics studies have demonstrated two classes of metal binding sites (high and low affinity). Both crystallographic and kinetic studies have typically used Mn^{2+} as the divalent cofactor. Kinetic experiments with $E_{1.7}$ (1.7 subunits) and Mn^{2+} show a tight-binding Mn^{2+} site with a $K_D = 5.0 \times 10^{-7}$ M and a weak-binding site with a $K_D = 4.5 \times 10^{-5}$ M per subunit.[41] Another set of measurements for E_{12} (all 12 subunits) yields dissociation constants of 5.3×10^{-7} M for the tight site and 1.7×10^{-5} M for the weak site, again with Mn^{2+} as the metal cofactor.[42] Less information is available for the magnesium-substituted enzyme, although it has been found that Mg^{2+} binds approximately 200-fold less tightly than Mn^{2+}.[43] A modified Hill plot has shown the half-saturation concentration of magnesium to be 0.47 mM, at which point 1.8 Mg^{2+} equivalents are bound.[44] This correlates well with crystallographic data for the Mn^{2+}-loaded enzyme.[40]

Although kinetic analyses have addressed several questions concerning cation binding and the possible functional roles of the two bound cofactors, most mechanistic insight has been derived from a careful analysis of structural data. Crystallographically it has been observed that the inner high-affinity metal site probably corresponds to the catalytic cofactor, while the weakly bound outer ion has been implicated with ATP binding.[40] Each of the two ions adopts octahedral coordination with an inter-ion distance of ~6 Å. By cocrystallization of a transition-state analog (L-methionine–S_R-sulfoximine), it was found that the analog binds close to the tight-binding metal ion, underscoring the catalytic role for that ion.

The chemical mechanism of glutamine synthetase involves initial formation of the γ-glutamyl phosphoryl intermediate by addition of the γ-phosphate from ATP to

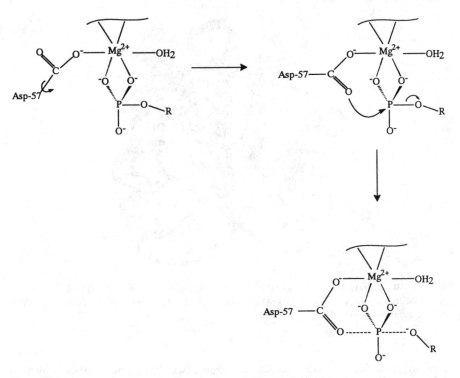

Figure 7.11   A proposed mechanism for *E. coli* Che Y. To initiate the reaction, one of the carboxylate oxygens on Asp-57 is believed to rotate toward the phosphate center. After rotation, nucleophilic attack occurs on the phosphate, yielding a pentacoordinate phosphate intermediate. This intermediate breaks down with RO^- as the leaving group. Magnesium is believed to orient the substrate and polarize the phosphate oxygens.[39]

the carboxylate carbon. Subsequent displacement of the activated phosphate by ammonia yields a tetrahedral transition-state complex. Mechanistic studies have been carried out but essentially provide overall details of the reaction pathway rather than insight on the specific bond-breaking and -forming steps. However, based on the pH-dependent behavior of glutamine synthetase, one mechanism has been proposed based on proton abstraction by a protein residue (Fig. 7.12).[45] The catalytic magnesium shown in Figure 7.12 serves two functions. First, the cation provides a template for the carbonyls and activates the phosphate toward transfer. After the phosphorylation step ammonia must be added to the activated carboxylate, and the magnesium ion stabilizes the transition-state complex mainly through electrostatic contacts.[45] These hypotheses have not yet been rigorously tested.

7.4.2   L-Aspartase

L-Aspartase catalyzes the reversible deamination of L-aspartic acid to form fumaric acid and ammonia. The *E. coli* enzyme is tetrameric with a total molecular weight of 192 kDa, and has an absolute requirement for divalent cations. Magnesium has a dissociation constant $K_d \sim 4.9 \pm 2.4$ μM.[46] The dissociation constants for Mn^{2+} and Co^{2+} are 0.8 ± 0.4 and 2.0 ± 0.3 μM, respectively.

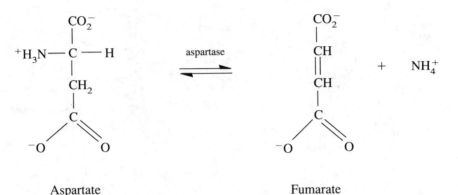

Aspartate   Fumarate

No inhibition has been reported at high concentrations of metal cofactor.[46] NMR relaxation studies indicate that a divalent metal ion must be bound under conditions of high pH to maintain activity.[46,47] At low pH the strict cation requirement disappears and residual activity occurs without the metal cofactor. The distinction between pH-dependent forms is linked to both divalent cation binding, and a residue with a pK_a value ~ 6.5 that has been tentatively assigned to a histidine (protonated for the amination reaction, and deprotonated for the deamination reaction).[47]

Since Mg^{2+} is spectroscopically silent, Mn^{2+} has been used extensively to probe the binding location and possible function of the metal cofactor. Through spin-lattice relaxation measurements, the distance between Mn^{2+} and the substrate analog O-phosphoserine was found to be 3.5–4.5 Å (depending on which oxygen is monitored).[46] This provides good evidence for cation binding at an allosteric or activator site since it is too far away to interact directly with substrate. However, this does not exclude the possible involvement of metal-bound waters, which can coordinate to the substrate through hydrogen bonds, as in the case of alkaline phosphatase (Fig. 7.8).

7.5 Photosynthesis and TCA Cycles

7.5.1 Ribulose-1,5-bisphosphate Carboxylase

RuBisco is one of the most abundant enzymes in nature, constituting up to 50% of all the soluble protein in leaves. The spinach enzyme is made up of eight large and eight small subunits (L_8S_8, where L is the catalytic subunit) with a total molecular weight of 550 kDa. Once activated it can catalyze either a carboxylation reaction in the photosynthetic carbon reduction cycle or oxidation of ribulose-1,5-bisphosphate

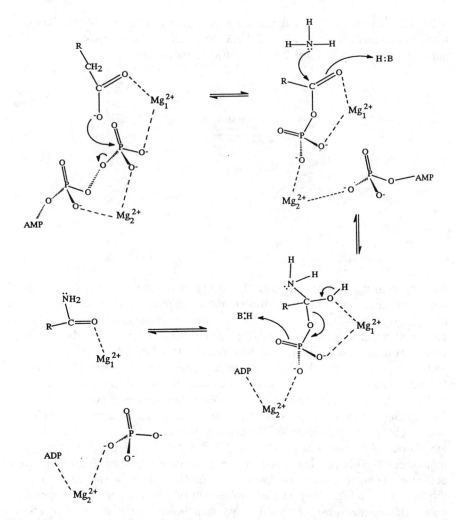

Figure 7.12 A proposed mechanism for *E. coli* glutamine synthetase amination involving the two bound magnesium ions. Two basic residues are thought to be involved in the reaction, supplying protons to the carboxyl and the phosphoryl groups. Ammonia attacks the susceptible carbonyl, eventually forming the peptide linkage. Other ligands to Mg^{2+} are not shown.[45]

in photorespiration.[48,49] A complete pathway is summarized in Scheme 7.1 for the carboxylation reaction,[49] and Table 7.3 lists all of the relevant binding constants. The final step in both reactions (see Scheme 7.1) is carbon–carbon bond cleavage with magnesium serving as a catalytic cofactor. Prior to both of these steps is an enolization reaction, where RuBisco isomerizes the substrate hydroxy ketone to yield the enediol form.

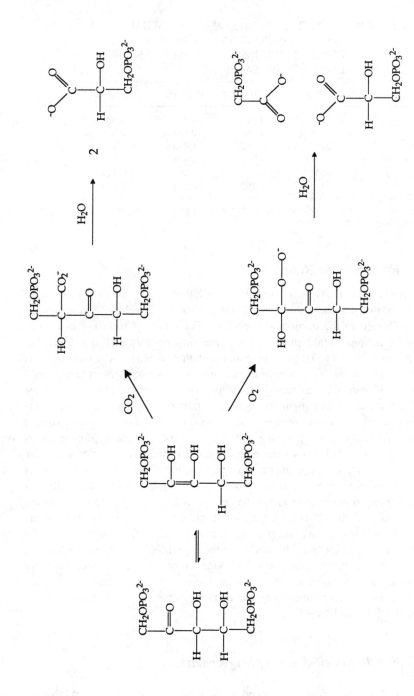

Scheme 7.1

Table 7.3 AFFINITY CONSTANTS AND HALF-METAL-
SATURATION CONSTANTS[a] FOR SOME
DIVALENT CATIONS AND RIBULOSE-1,5-
BISPHOSPHATE CARBOXYLASE[51]

Enzyme	Cation	K_a' (mM)	K_a (mM)	$K_{1/2}$ (mM)
Soya bean	Mn^{2+}	0.018	0.057	0.022
	Fe^{2+}	0.063	0.158	0.080
	Mg^{2+}	0.342	1.065	0.843
R. rubrum	Mn^{2+}	0.060	0.028	0.014
	Fe^{2+}	0.111	0.208	0.107
	Mg^{2+}	0.482	0.225	0.186

[a] K_a' is the observed metal binding constant, K_a is the apparent binding constant, and $K_{1/2}$ is the metal-ion concentration at half-maximal velocity.

7.5.1.1 Enolization Reaction

Considering only the first step here (recent crystallographic studies have focused on the inactive state of the enzyme),[52] it has been proposed that Mg^{2+} stabilizes the intermediate formed in the enolization reaction. For either of the oxygenation or carboxylation reactions to take place, the enzyme must be fully activated by carbamate formation (lysine + CO_2) and subsequent binding of Mg^{2+}. The carbamate is stabilized by Mg^{2+} through coordination of one of the carboxylate oxygens. Since the exact orientation of the substrate is unknown, two reasonable alternatives have been proposed for the subsequent involvement of magnesium in the enolization reaction. First, Mg^{2+} may polarize the C2-carbonyl, which not only serves to orient the substrate for enolization, but also stabilizes the oxygen anion after electron movement [Fig. 7.13(a,b)]. Polarization of oxygen also increases the acidity of the C3-H proton,[49] which is either abstracted by phosphate[49] or by a nearby lysine acting as a general base.[50] Second, it is possible that Mg^{2+} might polarize the C4 hydroxyl, assuming this orientation for the substrate [Fig. 7.13(c)] and, as before, increase the acidity of the C3-H hydrogen for abstraction by either Asn, His, or the carbamate.[49] In either capacity, magnesium serves to aid in the orientation of the substrate, and indirectly increase the acidity of the C3 proton for abstraction through polarization of a carbon–oxygen bond. The kinetics of enzyme activity have not been extensively studied. Binding affinities and kinetic parameters have been determined (Table 7.3), and no inhibition by the metal cofactor has been observed up to concentrations of 30 mM Mg^{2+}.

7.5.1.2 Synthesis of 3-phosphoglycerate

To date, most work has been concerned with activation of the protein, that is, formation of the ternary complex and the carboxylation step. The final step, small molecule synthesis, occurs with a bond-cleavage reaction, in this case yielding two molecules of 3-phosphoglycerate. The crystallographically characterized magne-

(a)

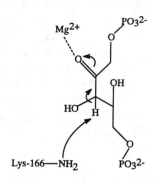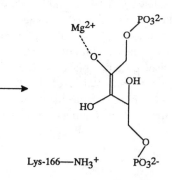

(b)

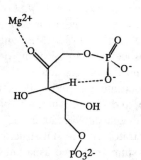

(c)

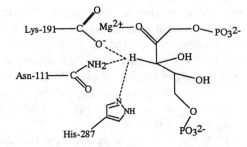

Figure 7.13 (a) A possible reaction sequence for *Rhodospirillum rubrum* ribulose-1,5-bisphosphate carboxylase using a specific substrate orientation where magnesium interacts with the C2-carbonyl. This scheme shows proton abstraction through Lys-166, while the cation polarizes the carbonyl. (b) The reaction sequence for the same substrate configuration as (a), but with proton abstraction by the C1 phosphate. (c) The reaction sequence where the substrate molecule is oriented with magnesium interacting with the C4-hydroxyl. Proton abstraction for this orientation is accomplished by one of the side chains of three possible residues: Lys-191, Asn-111, and His-287.[49,50,52]

sium cofactor in the spinach enzyme has a distorted octahedral coordination geometry with linkages to oxygens 2, 4, and 6 of the substrate analog 2-carboxyarabinitol bisphosphate (CABP), which contains a C3-OH instead of C3-H.[52] Details of the activation steps for bond cleavage remain unclear. Carbon–carbon bond cleavage in 1,5-ribulose bisphosphate occurs between carbons 2 and 3, which are also adjacent to the magnesium cofactor (see Scheme 7.1).[52] Although it is not known how C–C cleavage takes place in RuBisCO, it is thought that magnesium ion orients the substrates and stabilizes the 2,3-enediol intermediate. Heterolytic cleavage of the C2–C3 bond results in the localization of negative charge on C2, a conclusion supported by mutagenesis experiments, which show that formation of such an intermediate is sensitive to the position and environment of the Mg^{2+} cation.[52,53]

7.5.2 Isocitrate Dehydrogenase

Isocitrate dehydrogenase (IDH) catalyzes the oxidation of 2R,3S-isocitrate to yield α-ketoglutarate, using one equivalent of NAD^+. All known isocitrate dehydrogenases have an absolute requirement for a divalent cation, which binds at the substrate binding site [Fig. 7.14(b)].[54,55] Mechanistically the enzyme appears to oxidize substrate to oxalosuccinate, followed by decarboxylation and final stereospecific protonation to yield the α-keto acid.[56,57]

The substrate 2R,3S-isocitrate is bound at the active site in the *E. coli* enzyme by three arginines, one lysine, and a magnesium ion (Fig. 7.14).[54] Although the pocket is positively charged, the magnesium cofactor does appear to be essential for stabilization of the bound substrate. The crystal structure of the *E. coli* enzyme shows protein contacts to magnesium from two aspartates, two waters, and two oxygen ligands from the substrate.[54,58] The oxygen ligands from the substrate are the C2–OH, and one of the α-carboxylate oxygens.

Kinetic experiments have not yet been carried out to study Mg^{2+}-binding to IDH, or to probe metal-mediated catalysis. To date, isotope effects have shown that magnesium ion influences substrate binding (this is consistent with an isocitrate–Mg^{2+} complex as the true substrate) and the rate of decarboxylation.[57] It was also found that substrate binding was extremely metal ion dependent, again indicating that the true substrate is the Mg^{2+}-bound complex. Stopped-flow experiments on Mn^{2+}–isocitrate binding have demonstrated that this complex modulates the activity of $NADP^+$-IDH.[59] Recently an apparent $K(Mg^{2+}) = 125$ μM has been measured [compared to $K(Mn^{2+}) = 4.0$ μM].[60]

In what is believed to be the first step in the reaction, oxidation of isocitrate, a proton must first be removed from the C2 hydroxyl to form an oxygen anion.[58] The crystal structure shows Mg^{2+} coordinated to this hydroxyl,[54] and so it follows that magnesium is necessary for electrostatic stabilization of the alkoxide anion after proton loss.[58] Formation of the enolate (using the substrate analog α-ketoglutarate) is supported by ^{13}C NMR data.[61] In the second step of the reaction, after decarboxylation, magnesium gain appears to stabilize the resulting product anion.

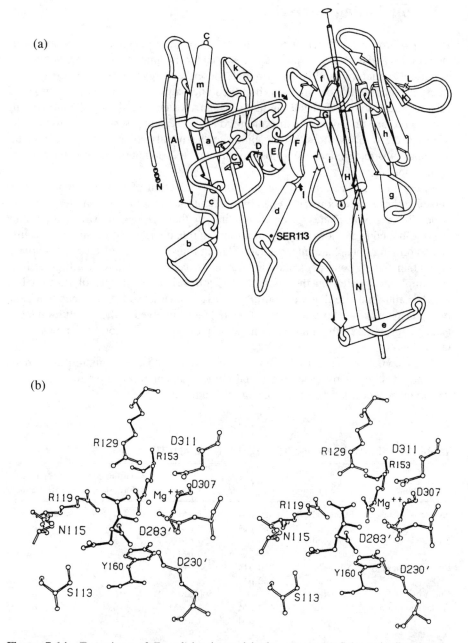

Figure 7.14   Two views of *E. coli* isocitrate dehydrogenase. (a) Schematic showing the α-helices and β-sheets of one monomer. I and II represent the front and rear pockets, while the shaded rod coincides with a twofold axis of symmetry. (b) Stereoview of the Mg^{2+}– isocitrate complex bound at the active site. Since the active site is at the interface, primed numbers indicate a distinct subunit. Isocitrate is shown with solid bounds. Adapted from ref. 58.

7.6 Concluding Remarks

Table 7.4 provides a succinct overview of the binding parameters associated with the proteins and enzymes discussed in this chapter. In spite of the apparent complexity of the structural and kinetic models described, several common themes do emerge. Notably, the enzymes in the glycolytic pathway described in Section 7.2 each require at least two metal binding sites: an allosteric regulatory site modulating either structure or binding, while the second ion typically serves a catalytic role. Also, in these cases the substrate binds to the metal-activated enzyme, and the principal substrate molecules (xylose, pyruvate, etc.) do not themselves have a high affinity for Mg^{2+}.

In contrast, in Section 7.3 we examined a selection of enzymes that catalyze phosphate-transfer chemistry. Typically divalent magnesium will complex to nucleotide di- or triphosphates prior to binding to the enzyme; however, additional enzyme-bound Mg^{2+} ions do play an important role in orienting substrate molecules and stabilizing phosphate intermediates during transfer. Many of these sites appear to contain two cations per catalytic subunit (except Che Y), with a high-affinity structural site located on the enzyme. By comparison to the glycolytic proteins, which maintain a site that was more allosteric in character, the enzymes that catalyze phosphoryl transfer often possess a site that is involved in the structural integrity of the protein. Both enolase and alkaline phosphatase are good examples of the latter.

The enzymes in Sections 7.4 and 7.5 tend to show isolated high-affinity metal binding sites. In many cases the metal cofactor apparently serves to stabilize an important reaction intermediate during the catalytic pathway.

Table 7.4 SUMMARY OF PROTEIN BINDING PARAMETERS

Enzyme	Type of Site	$K_{1/2}{}^{a,b}$	K_{Mg}	Coordination Before Substr.	Number After Substr.
Xylose isomerase	Structural (1)	80 μM	≤1 μM	4	6
	Catalytic (2)	80 μM	60 μM	6	6
Enolase	Structural	—	<20 μM	5	5
	Catalytic	—	50 μM	—	—
Pyruvate kinase	Allosteric	850 μM	750 μM	6	6
	Catalytic	850 μM	—	6	6
Che Y	Catalytic	—	500 μM	6	6
Glutamine synth.	Structural	470 μM	3.3 μM	6	6
	Catalytic	470 μM	1 μM	6	6
Aspartase		—	4.9 μM	—	—
RuBisCO		186 μM	482 μM	6	6
Isocitrate dehyd.		—	125 μM	—	6

[a] Metal concentration required to give half-maximal activity. Oringal references are noted in the text.

[b] Spaces with (—) indicate that the constant has not been determined.

References

1. Jenkins, J.; Janin, J.; Rey, F.; Chiadmi, M.; van Tilbeurgh, H.; Lasters, I.; De Maeyer, M.; Van Belle, D.; Wodak, S. J.; Lauwereys, M.; Stanssens, P.; Mrabet, N. T.; Snauwaert, J.; Matthyssens G.; Lambeir, A. M., *Biochemistry* **1992**, *31*, 5449–58.

2. Suekane, M.; Tamura, M.; Tomimura, C., *Agric. Biol. Chem.* **1978**, *42*, 909–17.

3. Rangarajan, M.; Hartley, B., *Biochem. J.* **1992**, *283*, 223–33.

4. Van Bastelaere, P.; Callens, M.; Vangrysperre, W.; Kersters-Hilderson, H., *Biochem. J.* **1992**, *286*, 729–35.

5. Lambeir, A.-M.; Lauwereys, M.; Stanssens, P.; Mrabet, N. T.; Snauwaert, J.; van Tilbeurgh, H.; Matthyssens, G.; Lasters, I.; De Maeyer, M.; Wodak, S. J.; Jenkins, J.; Chiadmi, M.; Janin, J., *Biochemistry* **1992**, *31*, 5459–66.

6. Black, C. B.; Huang, H.-W.; Cowan, J. A., *J. Coord. Chem. Rev.*, **1994**, in press.

7. (a) Jenkins, J.; Janin, J.; Rey, F.; Chiadmi, M.; van Tilbeurgh, H.; Lasters, I.; De Maeyer, M.; Van Belle, D.; Wodak, S. J.; Lauwereys, M; Stanssens, P.; Mrabet, N. T.; Snauwaert, J.; Matthyssens, G.; Lambeir, A. M., *Biochemistry* **1992**, *31*, 5449. (b) Van Tilbeurgh, H.; Jenkins, J.; Chiadmi, M.; Janin, J.; Wodak, S. J.; Mrabet, N. T.; Lambeir, A. M., *Biochemistry* **1992**, *31*, 5467. (c) Collyer, C. A.; Henrick, K.; Blow, D. M., *J. Mol. Biol.* **1990**, *212*, 211.

8. van Tilbeurgh, H.; Jenkins, J.; Chiadmi, M.; Janin, J.; Wodak, S. J.; Mrabet, N. T.; Lambeir, A.-M., *Biochemistry* **1992**, *31*, 5467–71.

9. Collyer, C. A.; Hendrick, K.; Blow, D. M., *J. Mol. Biol.* **1990**, *212*, 211–35.

10. Leiboda, L.; Stec, B., *Biochemistry* **1991**, *30*, 2817–22.

11. Faller, L. D.; Baroudy, B. M.; Johnson, A. M.; Ewall, R. X., *Biochemistry* **1977**, *16*, 3864–69.

12. Stubbe, J.; Abeles, R. H., *Biochemistry* **1980**, *19*, 5505–12.

13. Faller, L. D.; Johnson, A. M., *Proc. Natl. Acad. Sci. U.S.A.* **1974**, *71*, 1083–87.

14. Brewer, J. M.; Carreira, L. A.; Irwin, R. M.; Elliot, J. I., *J. Inorg. Biochem.* **1981**, *14*, 33–44.

15. Leiboda, L.; Stec, B.; Brewer, J. M., *J. Biol. Chem.* **1989**, *264*, 3685–93.

16. Wang, T.; Himoe, A., *J. Biol. Chem.* **1974**, *249*, 3895–902.

17. Anderson, V. E.; Weiss, P. M.; Cleland, W. W., *Biochemistry* **1984**, *23*, 2279–86.

18. Muirhead, H.; Clayden, D. A.; Cuffe, S. P.; Davies, C., *Biochem. Soc. Trans.* **1987**, *15*, 996–99.

19. Mildvan, A. S.; Cohn, M., *Abstracts of the Sixth International Congress of Biochemistry*, IUB Vol. 32, 1964, p. 322, IV-III.

20. Consler, T. G.; Uberbacher, E. C.; Bunick, G. J.; Liebman, M. N.; Lee, J. C., *J. Biol. Chem.* **1988**, *263*, 2794–801.

21. Mildvan, A. S.; Cohn, M. S., *J. Biol. Chem.* **1966**, *241*, 1178–93.

22. Mildvan, A. S.; Sloan, D. L.; Fung, C. H.; Gupta, R. K.; Melamud, E., *J. Biol. Chem.* **1976**, *251*, 2431–34.

23. Van Divender, J. M.; Grisham, C. M., *J. Biol. Chem.* **1985**, *260*, 14060–69.

24. Lin, M.; Turpin, D. H.; Plaxton, W. C., *Arch. Biochem. Biophys.* **1989**, *269*, 228–38.

25. Hassett, A.; Blattler, W.; Knowles, J. R., *Biochemistry* **1982**, *21*, 6335.

26. Oberfelder, R. W.; Barisas, B. G.; Lee, J. C., *Biochemistry* **1984**, *23*, 3822–26.

27. Kim, E. E.; Wyckoff, H. W., *J. Mol. Biol.* **1991**, *218*, 449–64.

28 Bosron, W. F.; Kennedy, F. S.; Vallee, B. L., *Biochemistry* **1975**, *14*, 2275–82.

29 Bosron, W. F.; Anderson, R. A.; Falk, M. C.; Kennedy, F. S.; Vallee, B. L., *Biochemistry* **1977**, *16*, 610–14.

30. Janeway, C.M.L.; Xu, X.; Murphy, J. E.; Chaidaroglou, A.; Kantrowitz, E., *Biochemistry* **1993**, *32*, 1601–9.

31. Xu, X.; Kantrowitz, E. R., *Biochemistry* **1993**, *32*, 10683–91.

32. Schlingting, I.; Almo, S. C.; Rapp, G.; Wilson, K.; Petratos, K.; Lentfer, A.; Wittinghofer, A.; Kabsch, W.; Pai, E. F.; Petsko, G. A.; Goody, R. S., *Nature* **1990**, *345*, 309–15.

33. Pai, E. F.; Krengel, U.; Petsko, G. A.; Goody, R. S.; Kabsch, W.; Wittinghofer, A., *EMBO J.* **1990**, *9*, 2351–59.

34. Foley, C. K.; Pedersen, L. G.; Charifson, P. S.; Darden, T. A.; Wittinghofer, A.; Pai, E. F.; Anderson, M. W., *Biochemistry* **1992**, *31*, 4951–59.

35. Schlingting, I.; Rapp, G.; Wittinghofer, A.; Pai, E. F.; Goody, R. S., *Proc. Natl. Acad. Sci. U.S.A.* **1989**, *86*, 7687–90.

36. Adari, H.; Lowy, D. R.; Willumsen, B. M.; Der, C. J.; McCormick, F., *Science* **1988**, *240*, 518–21.

37. Stock, A. M.; Mottonen, J. M.; Stock, J. B.; Schutt, C. E., *Nature* **1990**, *337*, 745–49.

38. Volz, K.; Matsummura, P., *J. Biol. Chem.* **1991**, *266*, 15511–19.

39. Stock, A. M.; Martinez-Hackert, E.; Rasmussen, B. F.; West, A. H.; Stock, J. B.; Ringe, D.; Petsko, G. A., *Biochemistry* **1993**, *32*, 13375–80.

40. Yamashita, M. M.; Almassey, R. J.; Janson, C. A.; Cascio, D.; Eisenberg, D., *J. Biol. Chem.* **1989**, *264*, 17681–90.

41. Villafranca, J. J.; Ash, D. E.; Wedler, F. C., *Biochemistry* **1976**, *15*, 536–43.

42. Hunt, J. B.; Ginsburg, A., *Biol. Chem.* **1980**, *255*, 590–94.

43. Denton, M. D.; Ginsburg, A., *Biochemistry* **1970**, *9*, 617–32.

44. Hunt, J. B.; Smyrniotis, P. Z.; Ginzburg, A.; Stadtman, E. R., *Arch. Biochem. Biophys.* **1975**, *166*, 102–24.

45. Colandrioni, J.; Nissan, R.; Villafranca, J. J., *J. Biol. Chem.* **1987**, *262*, 3037–43.

46. Falzone, C. J.; Karsten, W. E.; Conley, J. D.; Viola, R. E., *Biochemistry* **1988**, *27*, 9089–93.

47. Karsten, W. E.; Viola, R. E., *Arch. Biochem. Biophys.* **1991**, *287*, 60–67.

48. Lorimer, G. H.; Miziorko, H. M. *Biochemistry* **1980**, *19*, 5321–28.

49. Lundqvist, T.; Schneider, G., *J. Biol. Chem.* **1991**, *266*, 12604–11.

50. Lorimer, G. H.; Hartman, F. C., *J. Biol. Chem.* **1988**, *263*, 6468–71.

51. Christeller, J. T., *Biochem. J.* **1981**, *193*, 839–44.

52. Knight, S.; Andersson, I.; Branden, C.-I., *J. Mol. Biol.* **1990**, *215*, 113–60.

53. Lorimer, G. H.; Gutteridge, S.; Madden, M.-W., *Plant Molecular Biology,* von Wettstein, D.; Chua, N.-H. (Eds.), Plenum Press, N.Y., 1987, pp. 21–31.

54. Dean, A. M.; Koshland, Jr., D. E., *Nature* **1990**, *249*, 1044–46.

55. Hurley, J. H.; Dean, A. M.; Sohl, J. L.; Koshland, Jr., D. E.; Stroud, R. M., *Nature* **1990,** *249,* 1012–16.

56. Dean, A. M.; Koshland, Jr., D. E., *Biochemistry* **1993,** *32,* 9302–9.

57. O'Leary, M. H.; Limburg, J. A., *Biochemistry* **1977,** *16,* 1129–35.

58. (a) Hurley, J. H.; Thorsness, P. E.; Ramalingam, V.; Helmers, N. H.; Koshland, Jr., D. E.; Stroud, R. M., *Proc. Natl. Acad. Sci. U.S.A.* **1989,** *86,* 8635–39. (b) Hurley, J. H.; Dean, A. M.; Koshland, Jr., D. E.; Stroud, R. M., *Biochemistry* **1991,** *30,* 8671–78.

59. Farrell, Jr., H. M.; Deeney, J. R.; Hild, E. K.; Kumosinski, T. F., *J. Biol. Chem.* **1990,** *265,* 17637–43.

60. Muro-Pastor, M. I.; Florencio, F. J., *Eur. J. Biochem.* **1992,** *203,* 99–105.

61. Ehrlich, R. S.; Colman, R. F., *Biochemistry* **1987,** *26,* 2461–71.

8

Biological Chemistry of Magnesium Ion with Physiological Metabolites, Nucleic Acids, and Drug Molecules

J. A. Cowan

8.1 Introduction

Magnesium ion forms complexes with a variety of biological molecules, both macromolecules and smaller metabolites.[1-3] Table 8.1 illustrates the diversity of biochemical species that bind divalent magnesium, which in turn reflects the relatively high intracellular and extracellular concentrations of the ion in vivo. In earlier chapters we examined the chemistry of magnesium ion with proteins, enzymes, and ribozymes. In this chapter we will consider the coordination chemistry of magnesium with four classes of ligand: (1) low-molecular-weight substrates and metabolites; (2) the role of magnesium binding to antibiotics; (3) magnesium binding by ionophores; (4) the interaction of $Mg^{2+}(aq)$ with oligonucleotides.

Inasmuch as intracellular Mg^{2+} levels are relatively high, it is likely that magnesium complexes with such ligands are more common than previously thought. Some emphasis will be placed on magnesium-bound ligands that subsequently bind to nucleic acids and serve as antibiotics or antitumor agents by interfering with either the transcription or translation of genetic information. Under normal physiological conditions (typically millimolar), any molecule with an affinity $\geq 10^3$ M^{-1} will bind Mg^{2+}, and so many antibiotics are likely to be magnesium dependent, although the importance of this fact is not generally appreciated. In this chapter we will also build on the discussion developed in Chapters 4 and 5 since an understanding of metal–nucleic acid chemistry is relevant to an appreciation of this topic. In particular, we will examine crystallographically characterized magnesium-binding motifs associated with nucleic acids.

Table 8.1   CATEGORIZATION OF MAGNESIUM LIGANDS IN VIVO

Ligand	Function	Representative Examples
Enzymes	Structure stabilization, substrate activation, enzyme–substrate stabilization	Ribonuclease H, enolase, exonuclease, adenylate kinase
Proteins/peptides	Structural	Membrane surface
Hormones	Stabilization of oligomeric structure	Insulin
Antibiotics	Structural, stabilize complexes with proteins or nucleic acids	Mithramycin, adriamycin
Metabolites	Electroneutralization for transmembrane movement, substrate activation	Glucose phosphate, adenosine triphosphate
Ionophores	Transmembrane magnesium transport	A23187, ionomycin

8.2 General Metabolites

8.2.1 Phosphates

The most important magnesium-binding functional groups include phosphate, hydroxyl, and carboxylate. Later we shall see the use of 1,3-β-diketonate anions as a chelating unit; however, these are not used by common metabolic substrates. Both the relative importance and binding affinities of these sites follow the order: phosphates $>$ carboxylates, hydroxyls. This is firmly established by the data in Table 8.2. Association constants for simple monodentate ligands tend to fall in the range of 5–20 M^{-1}.[4-7] Comparison of the results for glucose phosphates, AMP^{2-}, and acetyl phosphate suggest that the hydroxyls on sugar units play a less important role in complexing Mg^{2+} than the phosphates. However, bidentate or tridentate coordination (for example, ADP^{3-} and ATP^{4-}) results in a significant increase in binding affinity.[8] For these reasons our attention will focus on the latter class, although note the use of magnesium ion in the context of enzyme reactions (Chapters 5 and 6) to serve as a template for group transfer or hydrolysis reactions (pyruvate kinase is a good example) of molecules that would not typically bind Mg^{2+} with high affinity in free solution.[9]

It is apparently essential that the ligand atoms to Mg^{2+} should carry at least a partial negative charge. Studies of the metal ion binding properties of dihydroxyacetone phosphate **1** and glycerol phosphate **2** show that in aqueous solution there is

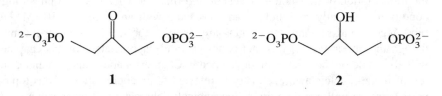

1   **2**

no interaction of the oxygen atoms at C-2 or C-3 of these ligands (either hydroxyl or carbonyl) with Mg^{2+}.[10] The stability of the complex depends only on the basicity of the phosphate. This can be contrasted with other ligands such as the pyrrolic

Table 8.2 MAGNESIUM BINDING TO PHOSPHATE, HYDROXYL, AND CARBOXYLATE LIGANDS

(a) Determination of kinetic and thermodynamic parameters for magnesium binding to phosphate-containing ligands by ^{25}Mg NMR[a]

Ligand	K_a (M^{-1})	k_{off} (s^{-1})	k_{on} (s^{-1})
[glucose-1-P]$^{2-}$	15	2.9×10^3	4.3×10^4
[glucose-6-P]$^{2-}$	8	3.1×10^3	2.5×10^4
$CH_3CO_2PO_3^{2-}$	9	1.5×10^3	1.4×10^4
AMP^{2-}	18	3.4×10^3	6.2×10^4
ADP^{3-}	2.2×10^3	2.5×10^3	5.5×10^6
ATP^{4-}	3.0×10^3	5.0×10^3	1.5×10^7

[a] Adapted from Ref. 21d. Data taken at 303 K. Refer to the original references for details of background salt levels (typically $\geq$ 100 mM Na$^+$ and/or Mg^{2+}).

(b) Metal binding to pyranose and furanose[a]

Cation	K (M^{-1}) pyranose	K (M^{-1}) furanose	Ionic strength (M)
Na$^+$	0.12	—	0.84–1.6
Mg^{2+}	0.19	—	1.15
Ca^{2+}	5.1	2.9	1.47
Sr^{2+}	5.1	1.9	1.32
Ba^{2+}	2.9	1.2	0.73
La^{3+}	10.4	8.7	0.90

[a] Adapted from Ref. 21d.

(c) Magnesium binding to carboxylates[a]

Ligand	K (M^{-1})	Ionic strength
Acetate	3.2	0.2 M KCl
Tartrate	40	0.2 M KCl
Malate	35	0.2 M KCl
Citrate	1585	0.16 M
Aspartate	269	0.1 M KCl

[a] Data from "Stability Constant," Special Publ. No. 17, Eds. A. Martell and L. G. Sillen, Chemical Society, London, 1964.

nitrogens in porphyrin (Fig. 1.3) or carbonyl oxygens in some of the antibacterial agents to be discussed later (Figs. 8.3–8.5), where a single negative charge is delocalized over several atoms. In these instances heteroatoms that bear even a partial negative charge may form an ionic bond to Mg^{2+}. Glycerol-1-phosphate takes on additional significance insofar as most membrane lipids are constructed from fatty acid esters of this molecule, with phosphate located as part of the charged head group (see Fig. 1.6).

8.2.2 Sugars

Ligation of alkali metal and alkaline earth ions to anionic sites of cell membranes is an important factor in understanding the dynamics of membrane chemistry,[11,12] impacting mechanisms of signal transduction and cell adhesion.[13,14] Other than the polyphosphate surface of lipid membranes, the hydroxyl functionality on sugars may serve as an important set of secondary binding sites. Table 8.2 notes binding constants of a variety of metal ions for glucopyranose **3** and fructofuranose **4**. Complexation with sugars may also play an important role in the recognition chem-

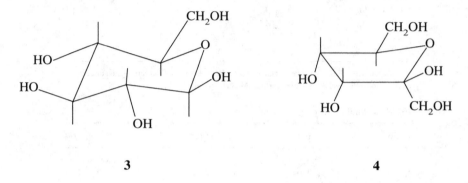

3   4

istry of glycoproteins.[15] Coordination of Mg^{2+} and Ca^{2+} by the sugar hydroxyls of a number of carbohydrate molecules has been demonstrated by crystallography,[16-18] and has been implicated in the transmission of nerve messages, changes in membrane structure, and bone formation.[19,20] Examples of complexes formed between sugar molecules and magnesium ion (or calcium) are summarized in Figure 8.1. Magnesium binds to two cis-vicinal hydroxyl groups of myo-inositol, retaining four H_2O ligands, which decreases the pucker of the sugar ring and lessens the 1,3-syn-diaxial steric interactions. Such conformational changes are also induced by calcium ion. One significant difference in the chemistry of magnesium versus calcium that has been alluded to earlier is the tendency of magnesium to retain its solvation sphere. As a result, while Mg^{2+} may bind to individual sugar units, it is Ca^{2+} that can cross-link several such units and is therefore a "structure-forming" cation. Such bridged structures have been identified for a number of sugars (for example, α-fucose). The differing solvation levels, and so coordination chemistry, reflect the smaller ionic radius of Mg^{2+} that consequently does not readily bind larger biological ligands. We shall see that the effect of this factor permeates all biological chemistry involving Mg^{2+} versus Ca^{2+}.

There have been few attempts to characterize the thermodynamic parameters associated with binding of alkaline earth metal cations to sugars.[21] Typically the systems studied have been phosphorylated at one or more hydroxyl positions. The impetus for such studies often derives from the importance of diphosphoinositide (DPI) and triphosphoinositide (TPI) in biological chemistry, with particular emphasis on its role in facilitating ion transport across membranes and stabilization of the myelin sheath.[22] In this regard, metal binding to trans-1,2-cyclohexanediol di-phosphate (CDP) **5** has been compared with glycerylphosphorylinositol diphosphate

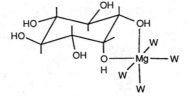

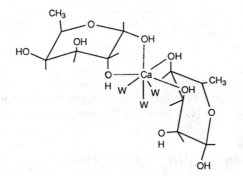

Figure 8.1 Magnesium binding to myo-inositol (top).[16] Two α-fucose molecules bind Ca^{2+} by chelation through pairs of hydroxyls (bottom).[18] Note the ability of Ca^{2+} to cross-link sugar molecules.

(GIP$_3$) **6** and glycerylphosphorylinositol phosphate (GIP$_2$) **6** (Table 8.3).[23] The distinct binding constants for GIP$_2$ and GIP$_3$ (reflecting DPI and TPI) suggests that

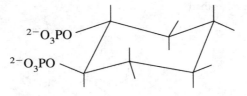

5

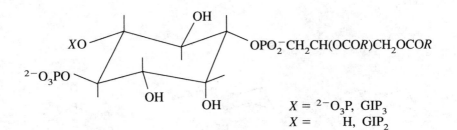

$X =$ $^{2-}O_3P$, GIP$_3$
$X =$ H, GIP$_2$

6

Table 8.3   BINDING CONSTANTS FOR PHOSPHOINOSITIDES AND RELATED LIGANDS

Cation[a]	CDP		GIP_3		GIP_2
	$\log K_{ML}$	$\log K_{MLH}$	$\log K_{ML}$	$\log K_{MLH}$	$\log K_{ML}$
Ca^{2+}	3.7	2.2	3.3	2.2	2.0
Mg^{2+}	3.7	2.3	3.5	2.4	2.3
Ni^{2+}	4.7	2.4	4.5	2.9	2.5

[a] CDP = 1,2-cyclohexanediphosphate; GIP_2 = glycerylphosphoryl inositol diphosphate; GIP_3 = glycerylphosphorylinositol triphosphate. Adapted from Ref. 11.

interconversion of such centers will effect the release or binding of metal ions. In turn this would alter the charge and stability of the surface membrane, and directly influence the transport of Na^+ or K^+ in the transmission of nerve impulses.

8.3 Magnesium-Binding Antibiotics

There exists a large number of antibacterial and antitumor agents that display a relatively high affinity for magnesium ion.[24] Each molecule possesses a chelating functionality that is typically based on one of the structural units illustrated in Figure 8.2. The significance of the divalent cation is often ignored in mechanistic discussions of drug function. However, it is becoming increasingly clear that the metal ion is an essential cofactor, rather than an adventitious appendage that plays a key role in the chemistry of the drug molecule. This will be clearly demonstrated by the examples to be described.

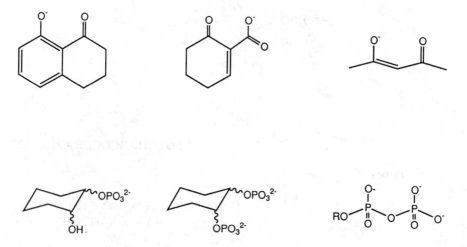

Figure 8.2   Summary of chelating functional groups on antibiotics and cellular metabolites.

8.3.1 Tetracycline

Tetracycline **7**, an antibacterial agent that inhibits bacterial protein synthesis by binding to the ribosome, chelates Mg^{2+} with a binding affinity of $2.5 \times 10^3 \, M^{-1}$ in a 10 mM tris buffer (pH 7.8, 25°C). The strongly fluorescent aromatic chromophore

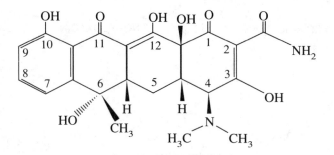

has been used to monitor binding to ribosomes where Mg^{2+} had been exchanged with Mn^{2+}.[25] The latter are paramagnetic and quench the drug fluorescence by increasing the rate of single–triplet intersystem crossing. Extracellular drug is likely to exist as either a Mg^{2+} or Ca^{2+} complex (inasmuch as the binding affinities of Mg^{2+} and Ca^{2+} are similar), while the extracellular $[Ca^{2+}]_e$ levels are slightly higher than $[Mg^{2+}]_e$ (Table 1.3). This may account for the ability of tetracyclines to inhibit the development of bone and teeth.[26,27] NMR evidence suggests that Mg^{2+} is chelated by the oxygen atoms on carbons 11 and 12.[28a] Since the drug contains three ionizable sites, analysis of the ionization levels is complex. Macroscopic pK_a values have been determined by analysis of NMR and potentiometric data.[28] The first ionization ($pK_1 \sim 3.33$) is usually attributed to the oxygen on carbon 3. The remaining two pKs (~ 7.75 and 9.61) are assigned to C12-OH and C10-OH. It is likely that pK_2 principally reflects ionization of C12-OH at the Mg^{2+} binding site.

8.3.2 Quinolone Antibiotics

Nalidixic acid was the first member of the quinolone family of antibacterial agents to be synthesized.[29] Several more potent derivatives have subsequently been developed (Fig. 8.3)[30]; however, all possess the latent ketocarboxyl chelating site that appears suitable for binding of divalent magnesium. Norfloxacin (Nor) is a well-studied quinolone antibacterial agent that binds to DNA in the presence of Mg^{2+}.[30–33] The likely magnesium binding site is shown in Figure 8.3. Unlike the aureolic acid family to be described, norfloxacin apparently binds as a monomer.[33] One effect of the magnesium-mediated binding to supercoiled DNA is a pronounced unwinding of the DNA.[31,32] Significantly, the maximal influence on DNA binding and unwinding arises with $[Mg^{2+}]$ close to physiological levels (ca. 2mM). Quinolone antibiotics bind most tightly to single-strand DNA, presumably promoted by π–π interactions of the planar aromatic ring and the DNA bases; however, it is

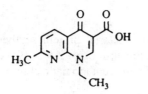

NALIDIXIC ACID

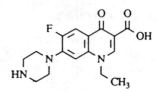

NORFLOXACIN

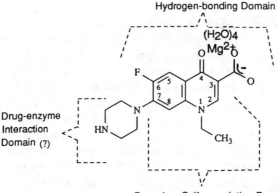

Figure 8.3 Nalidixic acid antibiotics and a schematic illustration of their likely mode of interaction with DNA and DNA binding proteins. Adapted in part from Ref. 32b.

unable to intercalate to double-strand DNA.[33] Binding is therefore favored by both the electrostatic influence of Mg^{2+} and favorable π–π interactions. Interestingly, single-strand nucleotides containing purine bases (A or G) bind the drug more effectively than pyrimidines, which may reflect the larger aromatic ring of the former.

Although the mechanism of action of this class of drugs has yet to be firmly established, a reasonable concensus model has the drug preferentially binding to single-strand DNA generated by the enzyme DNA gyrase. These drugs therefore inhibit gyrase activity, a key enzyme in DNA synthesis and replication. It is likely that Mg^{2+} plays an active role in DNA binding. With magnesium-binding affinities of $> 10^3$ M^{-1}, the complex will certainly be fully formed at typical in vivo [Mg^{2+}] levels. Binding models that neglect this fact may therefore have little physiological relevance. The Mg^{2+} ion neutralizes the repulsive interactions between the negatively charged carboxylate and the phosphate backbone, and may help to cross-link the two. Table 8.4 lists some other members of this drug family and DNA binding constants. In 10 mM tris/20 mM NaCl (pH 7.0, 25°C) the binding affinity of Mg^{2+} to Nor has been estimated as 2.5×10^4 M^{-1}, and that for binding of $Mg^{2+} \cdot$ Nor to DNA as 2.4×10^3 M^{-1}.[33]

8.3.3 Aureolic Acid Antibiotics

The aureolic acid family (Fig. 8.4) are effective against a wide selection of experimental and human tumors and function by inhibiting transcription of DNA as a result of complex formation in the presence of Mg^{2+}.[34] Mithramycin (MTC) is a member of this family of antibiotics, which includes the analogues chromomycin and olivomycin, and is characterized by attachment of two pyranose sugar chains and an aliphatic side chain to a central aureolic acid core.[35] NMR and optical spectroscopy have shown that these drugs bind as dimers with one Mg^{2+}(aq) cross-linking two mithramycin molecules by coordination to the C-1 carbonyls and C-9 hydroxyls (Fig. 8.4) and demonstrate high selectivity for 5'-XXGCXX-3' nucleotide sequences.[36-40] The identity of the bases at either side of the –GC– domain is relatively unimportant.[41,42] It has been previously suggested that the aureolic acid rings bind in the minor groove at 5'-GpC-3'. The longer sugar chains (rings C–E) also lie in the minor groove, while sugar rings A and B cross the phosphate backbone and lie in the vicinity of the major groove.[36-38] Mithramycin does not

Table 8.4　METAL BINDING CONSTANTS TO ANTIBIOTICS AND COMPLEXES WITH DNA

(a) Comparison of divalent cation complexes with chromomycin-d($A_2G_2C_2T_2$)

Parameter	Mg^{2+}	Ni^{2+}	Zn^{2+}	Cd^{2+}
Melting temperature (°C)	77	59	65	48
Half-life in water (h)	>63	Infinite	2.8	<0.25
Exchange time (Zn^{2+} for Mg^{2+}) (h)	61	—	—	—
Exchange time (Mg^{2+} for X^{2+}) (h)	—	Infinite	2.4	<0.25

Adapted from Ref. 45b.

(b) Stability constants for metal binding to daunorubicin[a]

Metal Ion	$\log K_M$	$\log K_7$[b]
H^+	10.0	—
Mg^{2+}	3.7	0.7
Ca^{2+}	3.3	0.3
Fe^{3+}	11.0	8.0
Zn^{2+}	4.5	1.5
Tb^{3+}	7.2	4.2

[a]　In aqueous solutions at 20°C and 0.15 M ionic strength. Adapted from Ref. 54.
[b]　Apparent stability constant at pH 7.

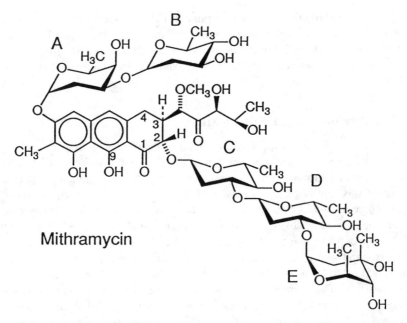

Figure 8.4 Aureolic acid antibiotics. The structure of mithramycin is shown, with the structures of olivomycin and chromomycin a_3 differing only in the conformations and modifications of the sugar rings.

intercalate nor unwind DNA, but does bind weakly to nucleotide sequences possessing an inverse 5'-*XXCGXX*-3' motif.[36] Drug action in the presence of a variety of metal cofactors has been examined. The drug–DNA complex could not be formed in the presence of Ca^{2+}, but could be stabilized by a variety of transition metal ions. In each case one divalent ion was required to form the dimer complex.[38b]

There is experimental evidence to support such a dimeric solution structure even in the absence of DNA.[40] The circular dichroism spectrum of an aqueous solution of mithramycin is shown in Figure 8.5. The feature at 280 nm arises from an absorption assigned to the naphthalenoid ring and is strongly polarized along the long axis.[41] Suillerot and co-workers have presented convincing evidence that mithramycin binds to DNA as a dimer in a right-handed screw conformation in the presence of Mg^{2+}, while at neutral or alkaline pH the left-handed conformation is adopted (Fig. 8.5).[42] At these pHs the C-9 phenolic hydroxyl is deprotonated, and so the molecule presumably adopts a left-handed conformation to reduce the electrostatic interactions between the negatively charged phenolates and may also serve to reduce the probability of steric clashes between the side chains around the aureolic acid moiety. Coordination by Mg^{2+} stabilizes the right-handed conformation, which also aligns the aureolic acid units and sugar chains for optimal binding to the DNA molecule.[43]

(a)

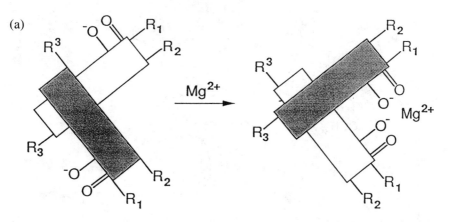

(b)

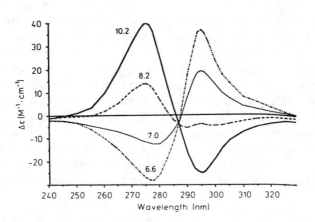

Figure 8.5 Mithramycin dimer and circular dichroism spectra. (a) In the neutral and metal-lated form the dimer exists in a right-handed screw conformation. The deprotonated dianion appears to adopt a left-handed conformation. (b) CD spectra obtained at various pHs showing the change in conformation in going from the neutral to dianionic species (adapted from Ref. 42).

8.3.4 Anthracycline Antibiotics

The anthracycline glycosides are a group of antibiotics thought to function by binding to DNA and inhibiting replication and translation.[28] There is a close structural similarity between daunomycin (daunorubicin) **8** and adriamycin (doxorubicin) **9.** Both contain the amino sugar daunosamine. This class of antibiotics has proven to

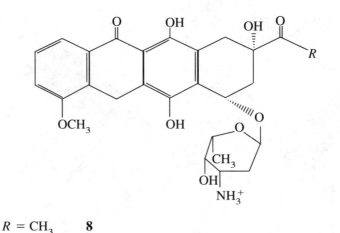

$R = CH_3$ **8**
$R = CH_2OH$ **9**

be highly effective in clinical use against both normal and neoplastic cells.[28] In fact, daunomycin was the first antibiotic drug to show activity against leukemia. Close inspection of the structural characteristics of a variety of drug molecules would lead one to suspect that many ought to bind Mg^{2+} with at least moderate affinity. Recent structural studies on adriamycin confirm this hypothesis,[44] with crystallographic evidence that the anthracycline ring intercalates between the DNA base pairs. A magnesium ion bridges the drug and DNA to stabilize the complex further.

Stability constants for metal binding to daunomycin are listed in Table 8.4(b). Daunomycin (and related drugs) also have a tendency to self-aggregate in solution, and an equilibrium constant of ~ 10 mM^{-1} has been estimated for dimerization.[45] Visual inspection of the structure shows two potential metal binding sites. At high metal concentration binding of a second cation has been found with Mg^{2+}, Ca^{2+}, and Tb^{3+}.[46]

8.4 Ionophores

8.4.1 Macrocyclic and Chelating Ethers

Ionophores are nonprotein ligands that carry metal ions across the hydrophobic lipid membrane barrier. The structures of several illustrative examples are shown in Figure 8.6.[47,48] Typically they possess polar functional groups to lend water solubility to the metal-free complex. After binding a metal ion these functional groups fold inward, leaving a hydrophobic surface that promotes solubility in the lipid membrane. Ionophores typically bind alkali and alkaline earth metals, of which the intracellular and extracellular concentrations are very different (Table 1.3), and so the maintenance of transmembrane concentration gradients by active transport processes is vital to normal cell metabolism. Since ionophores relax these concentra-

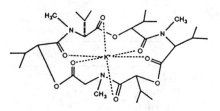

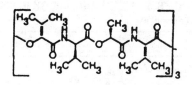

enniatin   valinomycin

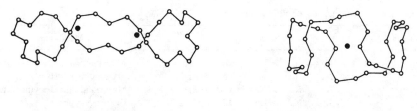

1:2 complex with Ba²⁺   1:1 complex with K⁺

Figure 8.6 Two potassium-selective cyclic peptide ionophores (enniatin and the structurally related valinomycin). The backbone structures below show the conformations adopted by valinomycin after formation of 1:2 and 1:1 complexes with Ba²⁺ and K⁺, respectively.

tion gradients, most ionophores are antibiotics (i.e., they result in a breakdown in cell metabolism that leads to cell death)..

Recent interest in ionophores has, in great measure, been promoted by the discovery that polyether ligands (especially macrocyclic ligands) have the ability to coordinate alkali and alkaline earth metal cations.[49] Subsequent studies have generated a wealth of coordination chemistry on a variety of synthetic ligand systems.[50-52] It therefore should come as no surprise that nature would utilize similar structural motifs to bind and transport Na⁺, K⁺, Mg²⁺, and Ca²⁺. Most of the ionophore systems that have been characterized actually show a high degree of selectivity for one or other of the alkali metals. Examples are documented in Table 8.5. Selectivity is based primarily on considerations of ion charge and size, and can be understood partly in terms of a "best-fit" criterion inasmuch as the ligands bind metal ions optimally when the ionic radii best match the "hole size" in the macrocycle. Note, however, the evidence that transport rates depend on lipid composition, and the dependence of selectivity on the type of cellular membrane, as clearly demonstrated by the data in Table 8.5(b). Ionophores also distinguish metal ions on the basis of hydration energies, since ions must lose their solvation shells before

Table 8.5 METAL UPTAKE BY IONOPHORES

(a) Selectivity

Ionophore	Specificity	Ionophore	Specificity
Macrocyclic ethers		Carboxylates	
Nonactin	$K^+ > Rb^+ > Na^+ > Ba^{2+}$	Nigericin	$K^+ > Na^+$
Monactin	$K^+ > Rb^+ > Na^+ > Ba^{2+}$	Monesin	$Na^+ > K^+$
Dinactin	$K^+ > Rb^+ > Na^+ > Ba^{2+}$	X-537	$Ba^{2+} > Sr^{2+} > Mg^{2+}$ $> Ca^{2+} > Mn^{2+} >$ $K^+ > Rb^+ > Na^+$
		A23187	$Ca^{2+} > Mg^{2+}$
Peptides			
Valinomycin	$Rb^+ > K^+ > Cs^+ > Ba^{2+} > Na^+$		
Eniatin B	$K^+ \sim Ba^{2+} > Rb^+ > Na^+ > Cs^+$		
Beauvericin	$Ca^{2+} \gg K^+ \sim Rb^+ \sim Cs^+ > Na^+$		

(b) Dependence of selectivities on the location of the cell membrane

Ionophore	Selectivity	Membrane
Monensin	$Na^+ > K^+ > Li^+ > Rb^+ > Cs^+$	Mitochondria and erythrocytes
	$Na^+ > Li^+ > K^+ > Rb^+ > Cs^+$	Erythrocyte ghosts
Nigericin	$K^+ > Rb^+ > Na^+ > Cs^+ > Li^+$	Mitochondria and erythrocytes
	$K^+ > Na^+ > Rb^+ > Li^+ > Cs^+$	Erythrocyte ghosts
X-206	$K^+ > Rb^+, Na^+ > Cs^+ > Li^+$	Erythrocyte ghosts
	$K^+ > Na^+$	Erythrocytes
Lasalocid	$Cs^+ > Rb^+ > K^+ > Na^+ > Li^+$	Mitochondria
	$Cs^+, Rb^+ > Na^+, K^+ > Li^+$	Erythrocyte ghosts
	$K^+ > Na^+$	Erythrocytes
Dianemycin	$Na^+, K^+, Rb^+, Li^+, Cs^+$	Mitochondria and erythrocytes
	$Li^+ > Na^+ > Rb^+, K^+ > Cs^+$	Erythrocyte ghosts
	$Na^+ \geq K^+$	Erythrocytes

Adapted from Ref. 48.

binding to the ligand. The charge–radius (q^2/r) ratio is higher for smaller ions (see Table 1.5), and so the energetics of solvent removal is an important aspect of ion-selective uptake. The ligand atoms must be able to coordinate to the metal center to make up for the loss of hydration energy, and so there is a tradeoff between solvent loss and ligand binding. Although most cations lose their entire solvation shell after ligation by the ionophore, a few examples are known where a metal cation binds more strongly than expected based on comparisons with ions immediately above and below in the periodic table. In these cases partial solvation can give a larger effective ionic radius.

Fewer examples of ionophores are known for the alkaline earths. Interestingly, many ionophores designed to bind M^+ ions as monomeric complexes show more elaborate structure chemistries with M^{2+} ions (Fig. 8.6).[47] These structural forms are extremely dependent on ionic radius. The chemistry of the valinomycin K^+-transporter is a good example (Fig. 8.6). The conformations of these drug molecules show a high degree of flexibility, and the preferred structure generally reflects the ionic radius and coordination number of a specific cation. Valinomycin is optimal for K^+. Smaller ions tend to be located at the periphery of the ligand. At low Mg^{2+} and Ca^{2+} concentrations a 2:1 (cation sandwich) structure is adopted.[53] Larger Ba^{2+} and Sr^{2+} ions actually prefer a 1:2 (peptide sandwich) structure, as shown in Figure 8.6.[47]

8.4.2 Carboxylic Ionophores

Alkaline earth ions are typically bound by carboxyl-bearing ionophores.[48] The negatively charged carboxylate helps to neutralize the divalent cation. Figure 8.7 illustrates the structures, and Tables 8.5 and 8.6 list the ion selectivities and binding affinities for a large number of carboxylic ionophores. This interesting class of ionophores can be used to couple metal uptake with proton release. For example, the ionophore A23187 possesses carboxylate ($-CO_2H$) functionality and can exchange H^+ for Mg^{2+} or Ca^{2+} to form salts. In this way ion gradients can be coupled.

Usually the ionophores favor 1:1 complex formation with monovalent ions. For divalent ions, a 1:2 metal ligand complex is formed unless the ligand has two ionizable groups. For example, ionomycin contains an additional β-diketone that yields an additional ionizable functional unit, and so a 1:1 complex is formed. Figure 8.8 also shows the 1:2 Ca^{2+}:A23187 structure.

As noted previously, coordination of a ligand to the divalent metal requires the replacement of solvent water molecules with ligand atoms. Desolvation is likely to arise by consecutive stepwise replacements (complete solvent removal in one step would be energetically difficult), and evidence points to a dissociative interchange pathway where each solvent water dissociates just prior to binding of a ligand atom.[54,55] Binding rate constants for Mg^{2+}, Ni^{2+}, and Mn^{2+}, for lasalocid and A23187 are consistent with rates of solvent exchange. This also suggests that any necessary ligand conformational changes are rapid and not rate limiting.

Selectivity in binding by ionophores has been analyzed theoretically, utilizing crystallographic data as experimental input.[56] Ionophore A23187 (calcimycin) shows selectivity for divalent over monovalent ions. Crystal structures for the ligand and its 2:1 complexes with Mg^{2+} and Ca^{2+} are available.[57] In both cases the metal is bound by carboxylate oxygens, the aromatic nitrogen of the benzoxazole ring, and the carbonyl oxygen of the ketopyrrole ring (Fig. 8.8). Additional stability arises from two hydrogen bonds formed from the NH of the ketopyrrole ring to the metal-bound carboxylate oxygen of the other ligand. The Ca^{2+} complex also possesses an additional H_2O to give a seven-coordinate metal ion; however, the binding affinities for Mg^{2+} and Ca^{2+} are similar.

8.5 Magnesium Binding to Nucleic Acids

In this section we will review the structural environments commonly identified for magnesium binding to nucleic acids, with emphasis on crystallographically well-characterized examples. As a prelude, we note the distinction between the two

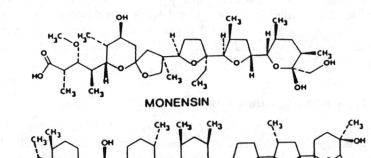

MONENSIN

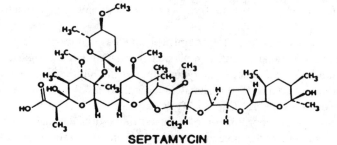

X-206

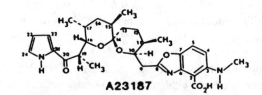

SEPTAMYCIN

A23187

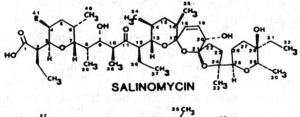

SALINOMYCIN

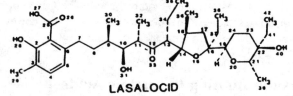

LASALOCID

Figure 8.7   Carboxylate ionophores. Adapted from Ref. 48.

Table 8.6 K_{app} VALUES FOR FORMATION OF 1:1 DIVALENT ION COMPLEXES BY MEMBRANE-BOUND A23187

Cation	Apparent complexation constant K_{app} (M^{-1})
Ca^{2+}	$(1.56 \pm 0.8) \times 10^5$
Mg^{2+}	$(6.58 \pm 0.4) \times 10^4$
Sr^{2+}	$(8.02 \pm 1.2) \times 10^3$
Ba^{2+}	$(4.83 \pm 1.7) \times 10^3$

Adapted from Ref. 58.

principal binding modes for divalent magnesium with a biological molecule. These are indirect (outer-sphere) association by hydrogen bond formation to solvent waters in the hydration sphere of the metal ion and direct (inner-sphere) coordination of a ligand to the metal center. Both modes of association have been identified in X-ray

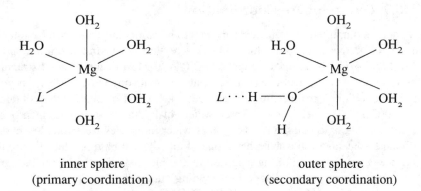

inner sphere
(primary coordination)

outer sphere
(secondary coordination)

crystallographic studies of metal–oligonucleotide complexes. Magnesium ion normally serves a structural role by stabilizing the sugar–phosphate backbone of oligonucleotides; however, a catalytic role has also been demonstrated for the reactions of ribozymes (see Chapter 5). The binding of positively charged counterions by the polyanionic sugar–phosphate backbone of nucleic acids is electrostatically favorable. Divalent magnesium alleviates electrostatic repulsive interactions between phosphates, thereby stabilizing base pairing and base stacking. This is evidenced by the increase in melting temperature (T_m) and hypochromism, respectively (Fig. 1.7). In contrast, transition metals that have a higher affinity for heteroatoms on the bases tend to inhibit base pairing and stacking, and so destabilize the double helix.[59,60] The stabilization of nucleic acid structure by alkali and alkaline earth ions is therefore of great structural significance. In this section we will review some characteristics of Mg^{2+} binding sites on oligonucleotides and illustrate these with specific examples.

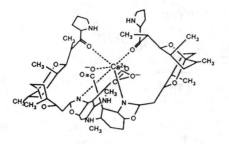
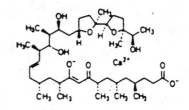

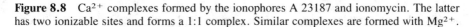

A 23187
2:1 complex with Ca²+

ionomycin
1:1 complex with Ca²+

Figure 8.8 Ca^{2+} complexes formed by the ionophores A 23187 and ionomycin. The latter has two ionizable sites and forms a 1:1 complex. Similar complexes are formed with Mg^{2+}.

8.5.1 Outer-Sphere Coordination

Outer-sphere binding to nucleic acids appears to be particularly common for interactions of the magnesium ion, which adopts a regular octahedral hydration sphere with a Mg^{2+}–H$_2$O bond distance close to 2.0 Å.[61] Most of the detailed structural work reported thus far has been carried out on small oligonucleotides. An octahedral hexahydrated magnesium ion has been identified on the Z-DNA obtained from a d(CGCGTG) sequence.[62] Three waters of hydration form hydrogen bonds with the wobble base pair G8T5. Two other water molecules were found to hydrogen bond to the sugar phosphate backbone of an adjacent base residue. Figure 8.9 shows the structural detail of the magnesium complex associated with the base pair GT and the corresponding magnesium binding site at the base pair G8C5 of the complementary Z-DNA oligomer d(CGCGCG).[63] In the latter case the hydrated magnesium ion moved somewhat closer to the purine residue, and only two Mg^{2+}-bound water molecules were able to hydrogen bond to the G8C5 base pair. The Mg(H$_2$O)$_6{}^{2+}$ ion forms an additional H bond to the O6 atom of G4 in the neighboring base pair G4C9.

A variety of data support outer-sphere coordination by Mg(H$_2$O)$_6{}^{2+}$ as the dominant binding mode for divalent magnesium. Stabilization of the hydration sphere can be ascribed to the formation of an extensive hydrogen-bond network with backbone phosphates, base heteroatoms, and sugar hydroxyls when bound to the major or minor grooves of ds DNA.[64-67] Given the rather small stability constant for a direct Mg^{2+}–phosphodiester complex ($K_a \sim 6$ M^{-1}; $\Delta G^0 \sim 1.1$ kcal mole^{-1}),[66,68] it is clear that this cannot compensate for energy lost from destruction of the H-bond network (~ 1–3 kcal mole^{-1} per bond), and so outer-sphere complex formation by Mg^{2+}(aq) is strongly favored on thermodynamic grounds.

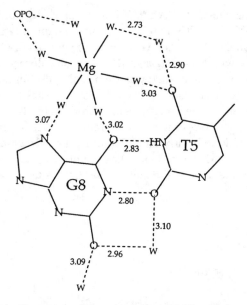

Figure 8.9 Schematic illustration of the outer-sphere Mg^{2+} complex with the wobble base pair GT in d(CGCGTG). Hydrogen bonds and distances are noted. Waters are indicated by W, and the backbone phosphate by OPO. Adapted from Ref. 1a.

8.5.2 Conformational Switching

Double-strand DNA normally adopts a B conformation; however, under certain solution conditions transitions to other conformations can be observed. The B → Z transition can be induced by high salt concentrations [either 2.5 M Na^+, or 0.7 M Mg^{2+}, or 0.04 M $Co(NH_3)_6^{3+}$]. Figures 8.9 to 11 illustrate some crystallographically characterized metal binding sites on Z-DNA. It should be noted that only certain sequences may adopt a Z conformation [e.g., poly(dG–dC), poly(dA–dC), poly(dT–dG)]. Alternating pyrimidine–purine bases are required. Ultimately the preferred backbone conformation is controlled by the stereochemistry of the ribose rings and orientations of base units that minimize steric and torsional strains.[69]

8.5.3 Inner-Sphere Coordination

The examples cited illustrate outer-sphere binding of magnesium ion to oligonucleotides by formation of a hydrogen-bond network from coordinated water molecules. However, direct inner-sphere binding has also been observed. Crystallographically characterized direct coordination sites for magnesium ions tend to lie at hairpin loops or the junction of two overlapping double-strand helices, where chelating sites may be formed by several backbone phosphate (for example, the strong Mg^{2+} binding sites in tRNA).[69-71] For example, the structure of the methylated hexamer d(m⁵CGTAm⁵CG) shows one magnesium ion octahedrally coordinated to five water molecules and one phosphate oxygen of G2.[72] It retains a regular octahedral geometry with $Mg-OH_2$ and Mg–phosphate distances of 2.0 Å. Figure 8.10 shows that two of the associated water molecules also form hydrogen bonds to neighboring backbone phosphates. Usually the favored sites for direct coordination to magnesium are phosphate oxygen and (surprisingly) N7 of guanosine. *Escherichia coli* tRNA[Phe] is one of the most thoroughly investigated RNA molecules. Crystallographic data have clearly identified at least four high-affinity magnesium binding sites ($K_a \sim 10^5$ M⁻¹)[73-76] and a large number of weaker sites ($K_a \sim 10^2$– 10^3 M⁻¹), although the structural details of the latter sites remain poorly characterized.[77,78]

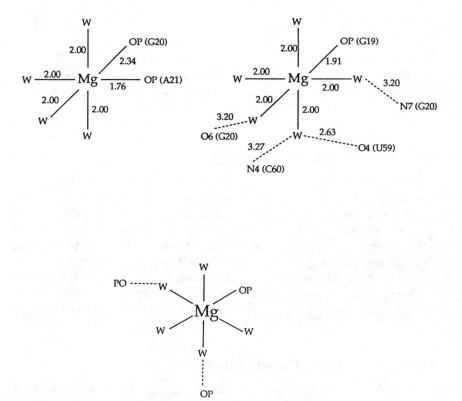

Figure 8.10 (Top) Two of the high-affinity Mg^{2+} sites crystallographically identified on tRNA[Phe], showing direct phosphate–magnesium contacts. (Bottom) A schematic diagram of one of the Mg^{2+} binding sites on Z-DNA d(m⁵CGTAm⁵CG). Hydrogen bonds from Mg^{2+}-bound solvent waters are indicated by dashed lines (see legend to Fig. 8.9). Both outer- and inner-sphere binding modes have been identified with the latter oligonucleotide, with the outer sphere as the most predominant form. Adapted from Ref. 1a.

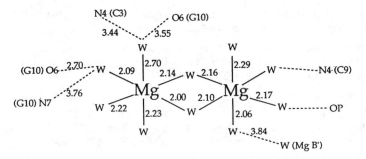

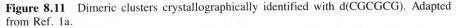

Figure 8.11 Dimeric clusters crystallographically identified with d(CGCGCG). Adapted from Ref. 1a.

8.5.4 Magnesium Clusters

Figure 8.11 shows two symmetry-related cation clusters that have been identified in the crystal structure of d(CGCGCG).[63] Each cluster contains two magnesium cations coordinated by water molecules in a slightly distorted octahedral environment. Magnesium ion typically adopts a regular octahedral geometry with a magnesium–oxygen distance of 2 Å. The two octahedrally coordinated ions share an edge through two bridging water molecules. The cluster on the convex side forms four H bonds to nucleotide bases, while the ion cluster on the groove side forms H bonds only to the phosphate backbone. Ion clusters (both two-ion and four-ion) have also been identified in the crystal structure of Z-DNA d(m⁵CGTAm⁵CG).[72,79]

8.6 Concluding Remarks

Divalent magnesium ion shows a rich coordination chemistry with low-molecular-weight metabolites, nucleic acid derivatives, and carrier ligands. Significant binding requires a chelating environment. Phosphate, carboxylate, hydroxyls, and ether oxygens are the predominant ligand groups. Both inner and outer (primary and secondary) coordination modes have been characterized. The latter is almost exclusively shown by complexes to oligonucleotides, since the major and minor grooves of double-strand nucleic acids have an array of phosphate and base heteroatoms in a geometric location that is appropriate for hydrogen bonding to Mg^{2+}-bound solvent waters.

Magnesium complexes of DNA-binding drugs are very common. In part the divalent metal ion electrostatically promotes complex formation with the polyanion, while waters of coordination may also contribute to the binding interactions.

References

1. (a) Black, C. B.; Huang, H.-W.; Cowan, J. A., *Coordn. Chem. Rev.* **1994,** in press. (b) Poonia, N. S.; Bajaj, A. V., *Chem. Rev.* **1979,** *79,* 389.

2. (a) Moore, P. B., *Nature* **1988**, *331*, 223–27. (b) Pyle, A. M., *Science* **1993**, *261*, 709. (c) Ryan, P. C.; Draper, D. E., *Biochemistry* **1989**, *28*, 9949–56.

3. (a) *Metals in Biology*, Vol. 26 (Ed. Sigel, H.), Marcel Dekker, NY, 1990. (b) Hughes, M. N., in *Comprehensive Coordination Chemistry* (Eds. Wilkinson, G.; Gaillard, R. D.; McCleverty, J. A.), Chap. 62.1, Vol. 6, Pergamon, 1987, pp. 541–754.

4. (a) Cowan, J. A., *Comments on Inorganic Chemistry* **1992**, *13*, 293–312. (b) Cowan, J. A., *Inorg. Chem.* **1991**, *30*, 2740–47.

5. (a) Bock, R. M., in *The Enzymes* (Boyer, P. D.; Lardy, H.; Kyrback, K., Eds.); 1960, pp. 3–38. (b) Phillips, R., *Chem. Soc. Rev.* **1966**, *66*, 501–27. (c) Izatt, R. M.; Christensen, J. J.; Rytting, J. H., *Chem. Rev.* **1971**, *71,*, 439–81.

6. (a) Kluger, R.; Wasserstein, P., *Biochemistry* **1972**, *11*, 1544–46. (b) Kluger, R.; Wasserstein, P.; Nakaoka, K., *J. Am. Chem. Soc.* **1975**, *97*, 4298.

7. (a) Oestreich, C. H.; Jones, M. M., *Biochemistry* **1967**, *6*, 1515–19. (b) Oestrich, C. H.; Jones, M. M., *Biochemistry* **1966**, *5*, 2926–31.

8. (a) Melchior, N. C., *J. Biol. Chem.* **1954**, *208*, 615–27. (b) Epp, A.; Ramasarma, T.; Wetter, L. R., *J. Am. Chem. Soc.* **1958**, *80*, 724–27.

9. (a) Cohn, M.; Hughes, T. R., *J. Biol. Chem.* **1960**, *235*, 3250–53. (b) Cohn, M.; Hughes, T. R., *J. Biol. Chem.* **1962**, *237*, 176–81. (c) Jaffe, E. K.; Cohn, M., *Biochemistry* **1978**, *17*, 652–57.

10. Liang, G.; Chen, D.; Bastian, M.; Sigel, H., *J. Am. Chem. Soc.* **1992**, *114*, 7780–85.

11. (a) Triggle, D. J., *Prog. Surf. Membr. Sci.* **1972**, *5*, 267–322. (b) Beveridge, T. J.; Murray, R.G.E., *J. Bacteriol.* **1976**, *127*, 1502–18. (c) Hendrickson, H. S.; Reinertsen, J. L., *Biochemistry* **1969**, *8*, 4855.

12. Ochiai, E. I., in *Biochemistry of the Elements*, Vol. 7, E. Frieden (Ed.), Plenum Press, New York, 1987, p. 205.

13. (a) Gaillard, T.; Hawthorne, J. N., *Biochim. Biophys. Acta* **1963**, *70*, 479. (b) Hauser, H.; Levine, B. A.; Williams, R.J.P., *Trends Biochem. Sci.* **1976,**, *1*, 278–81.

14. (a) Tschopp, J.; Jongeneel, C. V., in *Perspectives in Biochemistry*, Vol. II, Am. Chem. Soc., 1990. (b) Turner, R. S.; Burger, M. M., *Nature* **1973**, *244*, 509.

15. (a) Denman, J.; Marcel, M.; Bruyneel, E., *Biochem. Biophys. Acta* **1973**, *297*, 486. (b) Pricer, W. C.; Ashwell, G., *J. Biol. Chem.* **1971**, *246*, 4825.

16. Blank, G., *Acta Cryst.* **1973**, *B29*, 1677–83.

17. Cook, W. J.; Bugg, C. E., *J. Am. Chem. Soc.* **1973**, *95*, 6442–46.

18. Cook, W. J.; Bugg, C. E., *Biochim. Biophys. Acta* **1975**, *389*, 428–35.

19. Williams, R.J.P., *Chem. Soc. Quart. Rev.* **1970**, 331.

20. Cowan, J. A., *Inorganic Biochemistry. An Introduction.* VCH, New York, 1993, pp. 223–27.

21. (a) Bugg, C. E., *J. Am. Chem. Soc.* **1973**, *95*, 908. (b) Cook, W. J.; Bugg, C. E., *J. Am. Chem. Soc.* **1973**, *95*, 6442. (c) Rendleman, J. A., *Adv. Carbohydr. Chem.* **1966**, *21*, 209. (d) Angyal, S. J., *Aust. J. Chem.* **1972**, *25*, 1957.

22. Dawson, R.M.C.; Eichberg, J. *Biochem. J.* **1965**, *96*, 634.

23. Hendrickson, H. S.; Reinertson, J. L., *Biochemistry* **1969**, *8*, 4855–58.

24. *Metal Ions in Biological Systems* (Ed. Sigel, H.), Vol. 19, Antibiotics and their Complexes, Marcel Dekker, New York, 1985.

25. White, J. P.; Cantor, C. R., *J. Mol. Biol.* **1971**, *58*, 397.

26. Saxen, L., *Science* **1965**, *149*, 870.

27. Mull, M. M., *Amer. J. Dis. Child.* **1966**, *112*, 483.

28. (a) Martin, R. B., in *Metal Ions in Biological Systems* **1985**, *19*, 19–52. (b) Arcamone, F., *Doxorubicin*, Academic Press, New York, 1981.

29. Lesher, G. Y.; Froelich, E. J.; Garnett, M. D.; Bailey, J. H.; Brundage, R. P., *J. Med. Chem.* **1962**, *5*, 1063–65.

30. Shen, L. L.; Pernet, A. G., *Proc. Natl. Acad. Sci. USA* **1985**, *82*, 307–11.

31. Shen, L. L.; Kohlbrenner, W. E.; Weigl, D.; Baranowski, J., *J. Biol. Chem.* **1989**, *264*, 2973.

32. (a) Shen, L. L.; Baranowski, J.; Pernet, A. G., *Biochemistry* **1989**, *28*, 3879–85. (b) Shen, L. L.; O'Donnell, T. J.; Chu, D.W.T.; Cooper, C. S.; Rosen, T.; Pernet, A. G., *Biochemistry* **1989**, *28*, 3886–94.

33. Palu, G.; Valiseria, S.; Ciarrocchi, G.; Gatto, B.; Palumbo, M., *Proc. Natl. Acad. Sci. USA* **1992**, *89*, 9671.

34. Ward, D. C.; Reich, E.; Goldberg, I. H., *Science* **1965**, *149*, 1259–63.

35. Rao, K. V.; Cullen, W. P.; Sobin, B. A., *Antibiot. Chemother.* **1962**, *12*, 182–86.

36. (a) Banville, D. L.; Keniry, M. A.; Kam, M.; Shafer, R. H., *Biochemistry* **1990a**, *29*, 6521–34. (b) Banville, D. L.; Keniry, M. A.; Shafer, R. H., *Biochemistry* **1990b**, *29*, 9294–304.

37. Silva, D. J.; Kahne, D. E., *J. Am. Chem. Soc.* **1993**, *115*, 7962–70.

38. (a) Gao, X.; Patel, D. J., *Biochemistry* **1989**, *28*, 751–62. (b) Gao, X.; Patel, D. J., *Biochemistry* **1990**, *29*, 10940–56.

39. Berman, E.; Brown, S. C.; James, T. L.; Shafer, R. H., *Biochemistry* **1985**, *24*, 6887–93.

40. Demicheli, C.; Garnier-Suillerot, A., *Biochem. Biophys. Res. Commun.* **1991**, *177*, 511–17.

41. Cons, B.M.G.; Fox, K. R., *FEBS Lett.* **1990**, *264*, 100–4.

42. Demicheli, C.; Albertini, J.-P.; Garnier-Suillerot, A., *Eur. J. Biochem.* **1991**, *198*, 333–38.

43. (a) Sastry, M.; Patel, D. J., *Biochemistry* **1993**, *32*, 6558–604. (b) Keniry, M. A.; Banville, D. L.; Simmonds, P. M.; Shafer, R., *J. Mol. Biol.* **1993**, *231*, 753–67.

44. Cirilli, M.; Bachechi, F.; Ughetto, G.; Colonna, F. P.; Capobianco, M. L., *J. Mol. Biol.* **1992**, *230*, 878.

45. (a) Martin, S. R., *Biopolymers* **1980**, *19*, 713. (b) Chaires, J. B.; Dattagupta, N.; Crothers, D. M., *Biochemistry* **1982**, *21*, 3927. (c) Mitchell, P. R.; Sigel, H., *Eur. J. Biochem.* **1978**, *88*, 149.

46. Kiraly, R.; Martin, R. B., *Inorg. Chem. Acta* **1980**, *46*, 221.

47. Easwaran, K.R.K., in *Metals in Biology*, Vol. 26 (Ed. Sigel, H.), Marcel Dekker, NY, 1990, pp. 109–37.

48. Painter, G. R.; Pressman, B. C., in *Metals in Biology*, Vol. 26 (Ed. Sigel, H.), Marcel Dekker, NY, 1990, pp. 229–94.

49. Pedersen, C. J., *J. Am. Chem. Soc.* **1967**,, *89*, 7017.

50. Poonia, N. S.; Bajaj, A. Y., *Chem. Rev.* **1979**, *79*, 389.

51. Lehn, J. M., *Acc. Chem. Res.* **1978**, *11*, 149.

52. (a) Vogtle, F.; Weber, E., in *The Chemistry of Ethers, Crown Ethers, Hydroxyl Groups, and their Sulfur Analogues*, Part I (Ed. Patai, S.), Chap. 2, Wiley, New York, 1980. (b) Fenton, D. E., in *Comprehensive Coordination Chemistry* (Eds. Wilkinson, G.; Gaillard, R. D.; McCleverty, J. A.), Chap. 23, Pergamon, 1987, pp. 1–80.

53. Vishwanath, C. K.; Easwaran, K.R.K., *Biochemistry* **1982**, *21*, 2612.

54. Taylor, R. W.; Kauffman, R. F.; Pfeiffer, D. R., in *Polyether Antibiotics*, Vol. 1, (Ed. Westley, J. W.), Marcel Dekker, New York, 1983.

55. (a) Degani, H.; Friedman,, H. L., *Biochemistry* **1974**, *13*, 5022. (b) Shen, C.; Patel, D. J., *Proc. Natl. Acad. Sci. USA* **1976**, *73*, 4277.

56. Gresh, N.; Pullman, A., in *Metals in Biology*, Vol. 26 (Ed. Sigel H.), Marcel Dekker, NY, 1990, pp. 335–86.

57. Gresh, N.; Pullman, A., *Int. J. Quant. Chem. Quant. Biol. Symp.* **1983**, *10*, 215.

58. Kolber, M. A.; Haynes, D. H., *Biophys. J.* **1981**, *36*, 369.

59. A number of relevant articles are included in *Adv. Inorg. Biochem.*, Vol. 3, 103, 1981, (Eds. Eichorn, G. L.; Marzilli, L. A.).

60. A good summary may be found in *Metals and Micro-organisms*, Hughes, M. N.; Poole, R. K., Chapman and Hall, New York, 1989.

61. Agron, P. A.; Busing, W. R., *Acta. Crystallogr. C* **1985**, *41*, 8–10.

62. Ho, P. S.; Frederick, C. A.; Quigley, G. J.; van der Marel, G. A.; van Boom, J. H.; Wang, A. H.-J.; Rich, A., *EMBO J.* **1985**, *4*, 3617–23.

63. Gessner, R. V.; Quigley, G. J.; Wang, A. H.-J.; van der Marel, G. A.; van Boom, J. H.; Rich, A., *Biochemistry* **1985**, *24*, 237–40.

64. Cowan, J. A., *J. Am. Chem. Soc.* **1991**, *113*, 675–76.

65. Cowan, J. A., *Inorg. Chem.* **1991**, *30*, 2740–47; Reid, S. S.; Cowan, J. A.; *Biochemistry* **1990**, *29*, 6025–32.

66. Gessner, R. V.; Quigley, G. J.; Wang, A. H.-J.; van der Marel, G. A.; van Boom, J. H.; Rich, A., *Biochemistry* **1985**, *24*, 237–40.

67. Gessner, R. V.; Frederick, C. A.; Quigley, G. J.; Rich, A.; Wang, A. H.-J., *J. Biol. Chem.* **1989**, *264*, 7921–35.

68. (a) Stein, A.; Crothers, D. M., *Biochemistry* **1976**, *15*, 157. (b) Stein, A.; Crothers, D. M., *Biochemistry* **1976**, *15*, 160. (c) Stein, M. B.; Stein, A., *Biochemistry* **1976**, *15*, 3912.

69. Jack, A.; Ladner, J. E.; Rhodes, D.; Brown, R. S.; Klug, A., *J. Mol. Biol.* **1977**, *111*, 315–28.

70. Jack, A.; Ladner, J. E.; Klug, A., *J. Mol. Biol.* **1976**, *108*, 619–49.

71. Quigley, G. J.; Teeter, M. M.; Rich, A., *J. Mol. Biol.* **1978**, *75*, 64.

72. Wang, A. H.-J.; Hakoshima, T.; van der Marel, G. A.; van Boom, J. H.; Rich, A., *Cell* **1984**, *37*, 321–31.

73. Stout, C. D.; Mizuno, H.; Rao, S. T.; Swaminathan, P.; Rubin, J.; Brennan, T.; Sundaralingam, M., *Acta Crystallogr. B* **1978**, *34*, 1529–44.

74. Kim, S.-H., *Topics in Molecular and Structural Biology* **1981**, *1*, 83–112.

75. Holbrook, S. R.; Sussman, J. L.; Warrant, R. W.; Church, G. M.; Kim, S.-H., *Nucl. Acids Res.* **1977**, *8*, 2811–20.

76. Quigley, G. J.; Teeter, M. M.; Rich, A., *Proc. Natl. Acad. Sci. USA* **1978**, *75*, 64–68.

77. Jack, A.; Ladner, J. E.; Rhodes, D.; Brown, R. S.; Klug, A., *J. Mol. Biol.* **1977**, *111*, 315–28.

78. Reid, S. S.; Cowan, J. A., *Biochemistry* **1990**, *29*, 6025–32, and references therein.

79. Gessner, R. V.; Frederick, C. A.; Quigley, G. J.; Rich, A.; Wang, A. H.-J., *J. Biol. Chem.* **1989**, *264*, 7921–35.

CHAPTER 9

Genetics and Molecular Biology of Magnesium Transport Systems*

Ronald L. Smith and Michael E. Maguire

9.1 Introduction

Mg^{2+} is the most abundant divalent cation within the cells of both prokaryotic and eukaryotic organisms.[1,2] In contrast to the other biologically important cations, very little is known about its physiological roles within the cell. This paucity of knowledge can be attributed to the lack of adequate techniques for accurately measuring levels of intracellular Mg^{2+} or Mg^{2+} fluxes across biological membranes.[3-5] Mg^{2+} fluorescent dyes and microelectrodes lack the sensitivity and specificity required for routine quantitative analysis and are not suitable for all cell types. Radiochemical techniques for measuring unidirectional fluxes of Mg^{2+} are limited because of the prohibitive cost, unavailability, and short half-life of the $^{28}Mg^{2+}$ isotope.

Among the contributing factors to this lack of knowledge is the prevalent dogma, which implies that Mg^{2+} functions only as a stabilizing agent for membranes and proteins or as a cofactor with ATP in enzymatic reactions. However, recent studies have demonstrated that Mg^{2+} actively participates in the regulation of membrane channels and receptor proteins.[6,7] Furthermore, Mg^{2+} functions as a modulator for metabolic processes such as photosynthesis and oxidative phosphorylation (for reviews see Refs. 8 and 9). Many recent studies show significant changes in the intracellular concentration of Mg^{2+} and changes in Mg^{2+} flux in response to hor-

*RLS was supported by National Institutes of Health Training Grant T32-DK07319. Research from the laboratory of MEM is supported by National Institutes of Health Grant GM39447.

monal regulation.[10-13] These results have prompted several investigators to propose a regulatory role for intracellular Mg^{2+}.[1,14-19]

9.2 Chemical Properties of Mg^{2+}

A primary consideration in understanding these potential diverse physiological roles of Mg^{2+} is the mechanism by which the cation is translocated across the cell membrane. Mg^{2+} is the most charge dense of biologically relevant cations. This attribute results in a hydrated cation with a volume that is about three times greater than that of other common biological cations, and an atomic ion with a volume four times smaller. The bond strengths in the hydration shell are more than three orders of magnitude stronger than that of Ca^{2+} and four orders of magnitude stronger than those of Na^+ and K^+. Thus a Mg^{2+} transport system must initially interact with the hydrated cation, whose volume is about 350 times greater than the atomic ion that is eventually transported. By comparison, Ca^{2+} and Na^+ have hydrated cations whose volume is 25–30 times their volume of the atomic ions (Fig. 9.1). These inherent chemical properties suggest that transmembrane flux of Mg^{2+} may present a particular problem to the cell. Consequently, a Mg^{2+} transport protein may contain

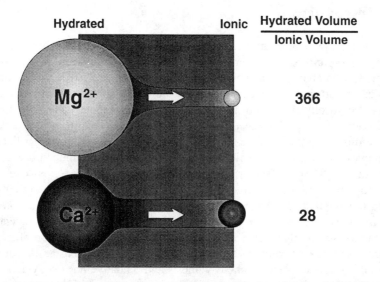

Figure 9.1 The problem of transmembrane flux. Cations in aqueous solution necessarily exist as the hydrated ion; however, for transport through a membrane, most if not all of this hydration shell is stripped off to expose the bare, atomic cation.[81] The figure compares, to scale, the hydrated Mg^{2+} and Ca^{2+} cations (left) and the corresponding atomic (unhydrated) cations (right) as they might traverse the membrane. As noted to the right of the figure, the ratio of volumes for the hydrated to unhydrated cation is >350 for Mg^{2+} and about 28 for Ca^{2+}. The volume ratio for Na^+ is similar to that of Ca^{2+}, while the corresponding ratio for K^+ is only about 4.

unique structural features that facilitate ligand binding and subsequent movement through the transport channel.

9.3 Mg^{2+} Transport in Eukaryotic Systems

Studies of Mg^{2+} transport proteins in eukaryotic systems are restricted to conventional biochemical assay and/or purification techniques. These have numerous inherent problems that are compounded by insensitive assay techniques. The lack of eukaryotic transport mutants prevents a genetic approach to investigation. A molecular route toward characterization is also impractical since extensive homologies with other eukaryotic transport proteins are only now being determined (see the following). This section briefly describes the phenotypic characteristics of the major eukaryotic Mg^{2+} transport systems thus far studied.

A Na$^+$-dependent Mg^{2+} efflux system has been described in chicken erythrocytes.[20] These studies measured Mg^{2+} efflux in preloaded cells and found that this activity followed Michaelis–Menten kinetics. This result implied a rate-limiting process that was associated with the cell membrane. The activity was determined to be energy dependent since an 80% reduction in net Mg^{2+} efflux could be achieved by treating with iodoacetamide and cyanide. Intra- and extracellular levels of Ca^{2+} had no effect on net efflux of Mg^{2+}. Efflux activity was found to be coupled with Na$^+$ influx, which occurred with a stoichiometric ratio of 2(Na$^+$):1(Mg^{2+}).[20] This Na$^+$/Mg^{2+} antiport activity was independent of existing Na$^+$/H$^+$ and Na$^+$/Ca^{2+} antiport systems. Although minor differences exist, similar systems have been reported in human erythrocytes and squid giant axons.[21-28] Thus an energy-dependent, Na$^+$-activated, Mg^{2+} efflux system that can be inhibited by amiloride, quinidine, and imipramine appears to be widespread.

Flatman and Smith have described a reversible sodium-dependent Mg^{2+} influx system in ferret erythrocytes.[29,30] This system mediates Mg^{2+} influx at high extracellular concentrations of Mg^{2+} and can be additionally stimulated by decreasing the exogenous Na$^+$ concentration. Elevated extracellular Na$^+$ concentrations reduce the Mg^{2+} influx activity. These properties suggest that Mg^{2+} influx is mediated by a Mg^{2+}/Na$^+$ antiport system. This system has also been shown to facilitate Mg^{2+} efflux.[29] Both influx and efflux activities are sensitive to inhibition by amiloride, quinidine, imipramine, and high extracellular levels of divalent cations. Neither influx nor efflux activity was significantly affected by ouabain or bumetanide. Vanadate was found to stimulate Mg^{2+} influx but not efflux. This system has some similarities to the Na$^+$/Mg^{2+} antiport activities described, but has sufficient distinct features that it likely involves different transport mechanisms.

Studies with S49 murine lymphoma cells identified a Mg^{2+} influx/efflux system that was not Na$^+$ dependent.[19,31-32] In these studies uptake of ^{28}Mg^{2+} occurred with a K_m of about 330 μM and a V_{max} of about 360 pmol/min/10^7 cells. Uptake via this system could be inhibited by β-adrenergic agonists, and stimulated by PGE and phorbol esters. Ca^{2+} and Mn^{2+} were poor and noncompetitive inhibitors while Ba^{2+} was a competitive inhibitor of influx.[10,33] Na$^+$ slightly stimulated influx of Mg^{2+}.

Mg^{2+} efflux activity was not affected by changes in the extracellular concentrations of Na^+, Ca^{2+}, or Mg^{2+}, nor was it affected by hormones. This result suggested that transport did not involve a Mg^{2+}–Mg^{2+} exchange mechanism and that influx and efflux were mediated by distinct transport systems.[19]

Romani and Scarpa have described a Mg^{2+} efflux pathway in cardiac and liver cells that appears to be hormonally regulated.[12] In these cells, efflux activity could be induced by β-adrenergic compounds such as norepinepherine and isoproterenol. Moreover, this efflux activity was sensitive to levels of cAMP. Efflux activity could be stimulated by forskolin and cAMP analogues, and repressed by vasopressin and carbachol. Moreover, low intracellular levels of cAMP significantly increased Mg^{2+} influx activity. Thus in cardiac and liver cells, cAMP appears to participate in the maintenance of Mg^{2+} homeostasis.

Quamme and Dai have described a Mg^{2+} influx system in Madin–Darby canine kidney (MDCK) cells.[34] This system was demonstrated to be specific for Mg^{2+} and independent of Ca^{2+} influx systems. Extracellular Ca^{2+} concentrations had no effect on the uptake of Mg^{2+} by this system. Likewise, Ca^{2+} uptake was not affected by the extracellular Mg^{2+}. This Mg^{2+} influx system could be completely and reversibly inhibited by La^{3+}, verapamil, and diltiazem with no corresponding effect on Ca^{2+} influx. Mg^{2+} influx was shown to be tightly controlled, with net uptake ceasing abruptly once cells obtained a basal level of free Mg^{2+}.

These studies illustrate the experimental limitations inherently associated with eukaryotic systems, primarily the inability to conduct detailed genetic and molecular characterizations. For these types of studies prokaryotic organisms are currently the obvious system of choice because of the extremely well-developed genetic systems and the relative ease of molecular manipulation. Moreover, the known gene families for prokaryotic transport systems have significant structural and mechanistic homology with their eukaryotic counterparts, thus allowing vertical transfer of the information gained.[35–39] Our laboratory has focused on the detailed molecular and genetic characterization of Mg^{2+} transport systems from the enteric bacteria *Salmonella typhimurium* and *Escherichia coli*. The results from these studies will be reviewed here.

9.4 Mg^{2+} Transport in Prokaryotes

9.4.1 Transport Systems of *Escherichia coli*

Mg^{2+} transport on prokaryotes was initially demonstrated in *E. coli* and *B. subtilis* by Silver[40,41] and Kennedy.[42–44] Silver and colleagues described an energy-dependent Mg^{2+} uptake system that was inhibited by the presence of other divalent cations. Mn^{2+} competitively inhibited Mg^{2+} uptake, while Ca^{2+} and Sr^{2+} had no detectable effect. Nelson and Kennedy subsequently demonstrated that Ni^{2+} and Co^{2+} could also inhibit Mg^{2+} uptake.[44] Co^{2+} was found to be a competitive inhibitor. Conversely, Mg^{2+} was found to be a competitive inhibitor of Co^{2+} uptake. This result suggested the hypothesis that Co^{2+} and Mg^{2+} uptake were facilitated by the same transport system. The cytotoxic effects of Co^{2+} were thus exploited in a selection for mutant strains unable to transport Mg^{2+}. Cells that were able to grow on a medium containing high levels of cobalt were isolated and presumed to lack the Co^{2+} uptake apparatus. These mutants (designed *cor* for *co*balt *r*esistance) were scored for their ability to uptake Mg^{2+}. Strains containing a *cor* mutation demonstrated a diminished capacity for uptake of both Co^{2+} and Mg^{2+} after growth in media supplemented with 10 mM Mg^{2+}. However in minimal growth medium with Mg^{2+} concentrations of about 25–100 μM, these strains exhibited a significant increase in Mg^{2+} uptake while still lacking detectable Co^{2+} uptake. This observation strongly suggested the existence of one or more additional independent Mg^{2+} transport systems. Park et al. confirmed that mutations at a single *corA* locus completely eliminated all uptake of $^{60}Co^{2+}$ and significantly decreased the amount of Mg^{2+} uptake.[45] Their studies also indicated that residual uptake of Mg^{2+} in a *corA* mutant strain could be repressed by supplementing the growth medium with Mg^{2+}. Park and colleagues mutagenized a *corA* mutant strain and scored for colonies that could not sustain growth on culture medium unless supplemented with 200 mM Mg^{2+}. Mutants with this phenotype were found to contain an additional lesion that was not linked to the *corA* locus. Genetic mapping experiments determined that the *corA* locus was located at 83 minutes on the *E. coli*[45] chromosomal map, while the second locus, designated *mgt*, was situated at 81 minutes. Both CorA and Mgt systems were similar with regard to Mg^{2+} affinity and velocity of uptake, although only the Mgt system was repressible by Mg^{2+}. The *corA mgt* double mutant still retained a small residual amount of repressible Mg^{2+} uptake, which pointed to a third as-yet-uncharacterized Mg^{2+} transport system.

9.4.2 Transport Systems of *Salmonella typhimurium*

Since *E. coli* and *S. typhimurium* are nearly identical organisms, it seemed likely that the Mg^{2+} transport systems would also be similar. This laboratory[46] isolated Mg^{2+} transport deficient mutants of *Salmonella typhimurium* by the cobalt resistance selection employed by Park et al.[45] These mutants lacked any detectable Co^{2+} uptake, exhibited marked cobalt resistance, and had a diminished capacity for Mg^{2+} uptake. The residual Mg^{2+} uptake could be further repressed by the addition of exogenous Mg^{2+} to the growth medium or greatly induced by omitting Mg^{2+}. These phenotypic data suggested that *S. typhimurium* also possessed at least two Mg^{2+} transport systems. Therefore *corA* strains were mutagenized with diethylsulfate and screened for colonies that could not grow on minimal growth medium unless supplemented with 100 mM Mg^{2+}. The resulting strains were found to require at least 50 mM Mg^{2+} supplementation to grow at wild-type rate. Mg^{2+} uptake in these strains was undetectable. Fine-structure genetic mapping indicated that this transport-deficient phenotype was the result of two separate mutations in addition to that at *corA*. The corA locus was mapped to 83.5 minutes on the chromosomal map of *S. typhimurium*. A second locus, designated *mgtA*, was mapped to 98 minutes, and a third locus, designated *mgtB*, was mapped to 80.5

chromosomal map units (Fig. 9.2).[47] Neither of these two *mgt* loci is situated in a map region comparable to the *mgt* locus of *E. coli*.

Previous studies of Mg^{2+} uptake in *E. coli* indicated that the *corA* locus could also mediate the efflux of Mg^{2+}.[48] This was confirmed for *S. typhimurium*.[46,49] Mutations at the *corA* locus resulted in the complete loss of Mg^{2+} efflux activity; this phenotype was independent of the presence or absence of the *MgtA* and *MgtB* alleles. Thus the CorA transport system appears to be the sole Mg^{2+} efflux system in *S. typhimurium*. A strain harboring mutations at all three transport loci (*corA*, *mgtA*, and *mgtB*) was constructed and tested for Mg^{2+} influx activity. The "triple mutant" strain could not grow unless supplemented with 100 mM Mg^{2+}, while a

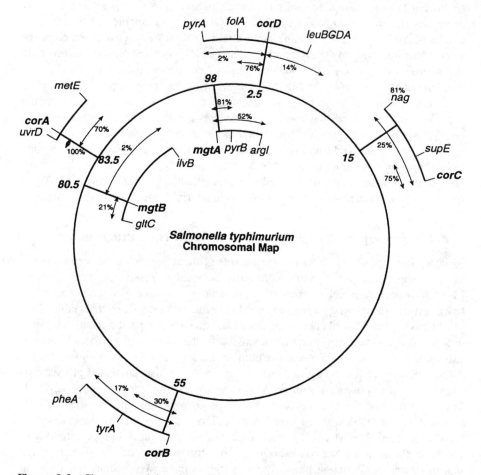

Figure 9.2 Chromosomal locations of the magnesium transport loci of *Salmonella typhimurium*. Chromosomal positions of loci encoding the three Mg^{2+} transport systems were determined by Hfr-directed conjugal matings and bacteriophage P22–mediated cotransduction. Bold numbers indicate the position in chromosomal map units (minutes), while cotransductional linkages with adjacent markers are expressed as a percentage.

wild-type copy of any one locus was sufficient to restore growth in unsupplemented medium (Fig. 9.3). Mg^{2+} uptake in the triple mutant was completely abolished.[47,49] Efflux activity could not be measured unless a wild-type *CorA* allele was supplied. These results indicate that the systems encoded by the *CorA*, *MgtA*, and *MgtB* loci are the only means of accumulating Mg^{2+} in *S. typhimurium*, and that the CorA system is the only system capable of mediating Mg^{2+} efflux.

9.4.3 Transport Properties of the CorA, MgtA, and MgtB Systems

In order to measure the kinetic properties of the three transport systems, a series of isogenic strains of *Salmonella typhimurium* were constructed by insertionally inactivating two of the three loci with transposable elements.[47,49] Each strain contained a wild-type allele of only one transport locus. Each of these strains was capable of growing in both minimal and complex media without additional Mg^{2+} supplementation. A strain containing insertion mutations at all three loci could not grow unless the growth media was supplemented with 100 mM Mg^{2+}.

9.4.3.1 The CorA Transport System

9.4.3.1.1 CorA-Mediated Mg^{2+} Influx

The transport system encoded by the *CorA* locus transports Mg^{2+}, Co^{2+}, Ni^{2+}, and possibly Mn^{2+}, with the highest affinity for Mg^{2+} and Co^{2+}. CorA facilitates the influx of Mg^{2+} with a K_m of about 15 μM at 20°C and a V_{max} of 250 pmol/min/10^8 cells. At 37°C the V_{max} is so high that it cannot be accurately measured, although it is estimated to be greater than 1500 pmol/min/10^8 cells. Co^{2+} uptake via the CorA system proceeds with a K_m of 30 μM at 20°C and a V_{max} of about 500 pmol/min/10^8 cells. Ni^{2+} is also transported by the CorA system with a K_m of 240 μM and a V_{max} of 360 pmol/min/10^8 cells (Table 9.1). As would be expected from their mutual transport by CorA, Mg^{2+}, Co^{2+}, and Ni^{2+}, each competitively inhibits the other's uptake. Mn^{2+} also competitively inhibits Mg^{2+} uptake, while Ca^{2+} has no effect on Mg^{2+} or Co^{2+} influx (Table 9.2).[49]

9.4.3.1.2 CorA-Mediated Mg^{2+} Efflux

Silver and Lusk reported that the CorA transport system of *E. coli* could also mediate Mg^{2+} efflux.[48] This observation was confirmed in *S. typhimurium*.[50] Mutations at the *CorA* locus completely abolish efflux activity regardless of whether the remaining two Mgt systems are present (Fig. 9.4). Efflux activity is virtually undetectable at extracellular Mg^{2+} concentrations of 20 μM, which is the K_m value for Mg^{2+} influx (Fig. 9.5). Efflux is "activated" at increasing extracellular Mg^{2+} concentrations. The CorA system extrudes Mg^{2+} at half the maximal efflux rate at an extracellular Mg^{2+} concentration of about 1 mM. Thus efflux appears to proceed by a gated Mg^{2+}–Mg^{2+} exchange mechanism in which efflux is activated by high levels of extracellular Mg^{2+}.

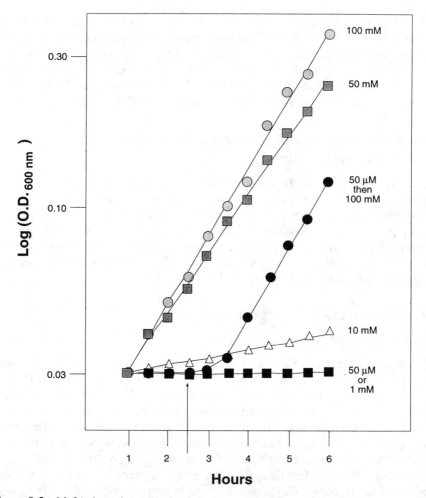

Figure 9.3 Mg^{2+}-dependent growth of the transport-deficient strain MM281. *Salmonella typhimurium* strain MM281 contains insertion mutations at all three Mg^{2+} transport loci (*corA*, *mgtA*, and *mgtB*). Cells were inoculated into Luria-Bertani (LB) broth supplemented with MgSO$_4$ at the indicated concentrations. Growth was scored as an increase in optical density at 600 nm. In one assay, cells were initially inoculated into LB containing 50 μM Mg^{2+}, and after 2.5 hours incubation the Mg^{2+} concentration was increased to 100 mM. Wild-type *S. typhimurium* LT2 can sustain growth in LB with no additional Mg^{2+} supplementation. Growth of the triple mutant strain, MM281, approaches the wild type rate of growth when the medium is supplemented with 100 mM Mg^{2+}. No growth is observed in unsupplemented LB. Strains that contain a single wild type loci grow in unsupplemented LB medium at a rate equal to that of MM281 in LB containing 100 mM Mg^{2+}.

Previous work by Park et al. described a second cobalt resistance locus in *E. coli*, which was designated *corB*.[45] Gibson et al. identified three additional loci in *S. typhimurium*, which conferred Co^{2+} resistance upon disruption.[50] The levels of resistance were significantly weaker than those of a *corA* mutant. A *corA* mutant is resistant to Co^{2+} concentrations in excess of 300 μM in broth culture, while *corB*,

Table 9.1 KINETIC PROPERTIES OF Mg^{2+} INFLUX IN
SALMONELLA TYPHIMURIUM[a]

Transport System	Cation	K_m (μM)	V_{max} (pmol min^{-1}/10^8 cells)
CorA	Mg^{2+}	15	250
	Co^{2+}	30	500
	Ni^{2+}	240	360
MgtA	Mg^{2+}	30	150
	Ni^{2+}	5	20
MgtB	Mg^{2+}	6	75
	Ni^{2+}	2	15

[a] Kinetic values were determined in strains possessing a single wild-type transport locus. Activities were measured after growth in medium containing exogenous Mg^{2+} at either 10 mM (CorA) or 5 μM (MgtA and MgtB). Uptake assays were performed as described by Grubbs et al. at 37°C.[80] Influx via the CorA system was measured at 20°C to bring uptake within a linear range. K_m values at 37°C cannot be estimated although they are not likely to differ significantly from those obtained at 20°C. The V_{max} for Mg^{2+} influx via CorA is estimated to be greater than 1000 pmol min^{-1}/10^8 cells.

corC, and *corD* mutants are only resistant to about 125 μM Co^{2+}. Fine-structure genetic mapping experiments indicated that the three loci were unlinked and situated at 55, 15, and 3 map units, respectively (Fig. 9.2).

The *corB*, *corC*, and *corD* mutations had nominal effects on CorA-mediated Mg^{2+} influx. Any single mutation decreased the V_{max} for Mg^{2+} influx by a factor of 2 with no effect on the apparent K_m. A *corBCD* triple mutant had no additional effect. Single *corB*, *corC*, and *corD* mutations were found to the reduce the half-maximal velocity of Mg^{2+} efflux by a factor of 3 when extracellular Mg^{2+} concentrations were at 10 mM. Mutations at all three loci resulted in complete abolition of Mg^{2+} efflux activity in media containing 10 mM Mg^{2+} (Fig. 9.4). These results indicate that Mg^{2+} efflux requires the gene product of *CorA* and any one of the *CorB, C,* or *D* loci.[50] This is in contrast to Mg^{2+} influx, which can be mediated by

Table 9.2 APPARENT K_i VALUES FOR CATION INHIBITION OF Mg^{2+} INFLUX IN
SALMONELLA TYPHIMURIUM[a]

Transport system	Ca^{2+} [b]	Co^{2+}	Mn^{2+} [c]	Ni^{2+}	Zn^{2+} [d]
CorA	5000	50	30	500	2
MgtA	300	40	partial	30	7
MgtB	>30,000	8	40	13	—

[a] The K_i values for cation inhibition of Mg^{2+} uptake were calculated from the concentrations that result in a 50% inhibition of influx. These concentrations include the exogenous Mg^{2+} concentrations and the K_a of the specific transporty system.[49]

[b] Ca^{2+} inhibits Mg^{2+} uptake by MgtB by less than 10% at 30 mM Ca^{2+}. Higher Ca^{2+} concentrations result in aggregation of cells.

[c] Mn^{2+} inhibits Mg^{2+} uptake by MgtA at a maximal value of 35%.

[d] Zn^{2+} does not significantly inhibit Mg^{2+} uptake by the CorA or MgtB systems at 100 μM. Concentrations greater than 100 μM result in immediate cell lysis.

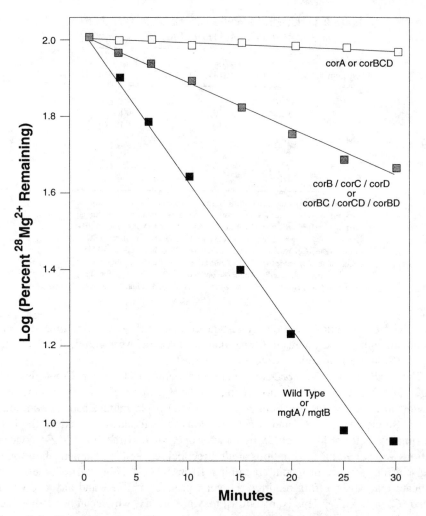

Figure 9.4 The effects of mutations at *cor* and *mgt* loci on Mg^{2+} efflux. $^{28}Mg^{2+}$ efflux was measured in strains of *S. typhimurium* possessing mutations at the indicated loci by the methods of Snavely et al.[49,57] Efflux in the wild type strain cannot be distinguished from strains harbouring mutations at the *MgtA* or *MgtB* loci. Mutations at the *CorA* locus eliminate detectable efflux activity. Mutations at any one or two of the *CorB*, *CorC*, or *CorD*, loci reduce the half-time for efflux by a factor of 3 while mutations at all three loci abolish efflux.[50]

the *CorA* gene product alone. It is not yet known whether the *CorBCD* gene products play an active role in cation translocation or indirectly modulate efflux activity through additional unidentified processes.

9.4.3.1.3 The *CorA* loci

Recombinant plasmids encoding reciprocally complementing *CorA* alleles from *E. coli* and *S. typhimurium* were constructed and used to identify the gene prod-

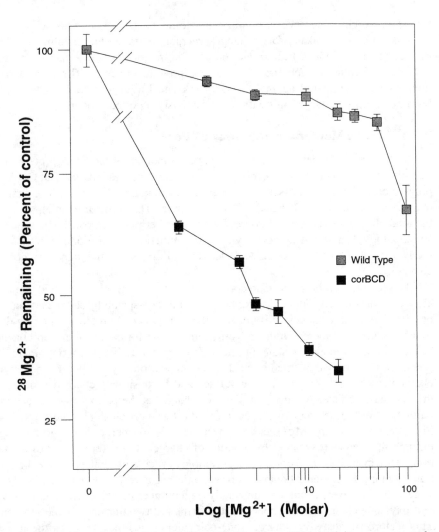

Figure 9.5 Effect of extracellular Mg^{2+} on Mg^{2+} efflux. Efflux activity was measured as described by a modification of the procedure described by Snavely et al.[49,57] Efflux activity was initiated by resuspending the washed cells in medium containing Mg^{2+} at the concentrations indicated. Mg^{2+} concentrations greater than 20 μM have little or no additional effect on the rate of efflux from the wild type strain. Control experiments indicate that the efflux rate is linear at any given extracellular Mg^{2+} concentration for a period of at least 30 minutes.[50]

ucts.[46,51] The *CorA* loci from both species encode a single polypeptide with gel molecular weights of 37 (*E. coli*) and 42 kDa (*S. typhimurium*). These proteins were determined to be integral membrane proteins by biochemical criteria.[51]

Sequence analysis of both alleles indicated that the *CorA* locus is comprised of a single open reading frame that encodes a polypeptide 316 amino acid residues in length with a predicted molecular mass of 37.6 (*E. coli*) and 36.5 (*S. typhimurium*) kDa. The deduced amino acid sequences for these two peptides are 98% identical.

Six of the eight differences are situated within the first 69 amino acids of the protein.[52] CorA is unusual for an integral membrane protein in that almost 29% of the amino acid residues bear frank charge. The majority of this charge is evenly distributed throughout the first 80% of the protein sequence. The predicted sequences were compared against the Genbank and EMBL databases and found to have no significant homology with any previously reported protein.

9.4.3.1.4 The Membrane Topology of CorA

Smith et al. recently determined the membrane topology of the CorA protein using well-described protein fusion techniques.[52–56] These studies indicate that CorA is a polytopic protein of two essentially independent domains (Fig. 9.6). The first domain contains the first 235 residues of the protein. This structure contains 88% of the total charge and is situated entirely within the periplasmic space. The sequence and topology data do not suggest any particular function for this domain, although it likely mediates the acquisition and delivery of ligand to the transport pore or channel.

The second, membrane domain consists of the remaining 81 amino acid residues. This domain contains three hydrophobic membrane-spanning segments that are predicted to form α-helices. It is unlikely that a monomeric protein containing three transmembrane segments would be sufficient for the formation of an ion channel. Therefore, the functional state of the CorA protein is likely an oligomer. The boundaries of individual membrane helices are reasonably defined by negatively charged residues at the periplasmic surface and by positively charged residues on the cytoplasmic face. A cluster of negative charges at the periplasmic surface may participate in the delivery of the cation to the transport channel. One might intuitively expect that the Mg^{2+} cation, because it is so charge dense, might move through the transport channel by a series of charge–charge (electrostatic) interactions. However, it is unlikely that CorA employs such a mechanism since the three transmembrane segments contain only a single negatively charged residue. These helices do, however, contain an unusually high number of amino acid residues with hydroxyl side chains that are concentrated in the last two transmembrane segments. This feature suggests that charge–lone-pair interactions may serve as the mechanism for cation movement. This hypothesis is currently under investigation.

9.4.3.2 The Mgt Transport Systems

9.4.3.2.1 Transport Properties of MgtA and MgtB

MgtA and MgtB are inducible Mg^{2+} transport systems that mediate influx only, during conditions of relative Mg^{2+} deprivation. Both systems are repressed under normal growth conditions at exogenous Mg^{2+} concentrations above about 50 μM.[49] Transcriptional analysis of mgtA::lacZ and mgtB::lacZ operon fusions indicate that repression occurs at gene level. Repression of the Mgt systems is relieved as extracellular Mg^{2+} concentration falls, with transcription increasing greatly below about 100 μM Mg^{2+}. At this concentration, MgtA and MgtB transcription have

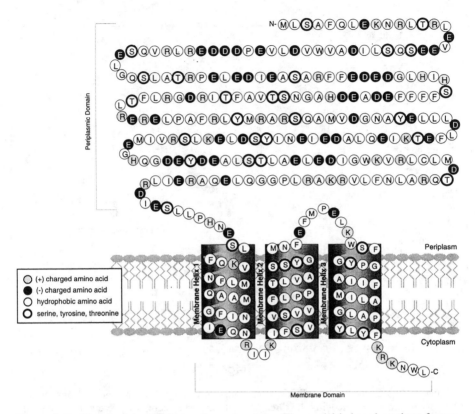

Figure 9.6 Model of the CorA membrane topology. The model is based on three factors, the enzymatic activities of BlaM and LacZ fusions, hydropathy predictions, and the distribution of positive charge. The boundaries of each membrane helix are approximations based on charge. Each residue is denoted by its single letter code. Positively charged residues are indicated by lightly hatched circles, negatively charged residues are indicated as black circles, and residues that bear hydroxyl side chains are represented by open circles with bold lines.[52]

increased by about tenfold. At 10 μM extracellular Mg^{2+} concentrations, transcription of *MgtB* but not *MgtA* continues to increase by a total of about 1000-fold (Fig. 9.7).[57]

Like CorA, the MgtA and MgtB systems transport both Mg^{2+} and Ni^{2+}, which are mutually competitive inhibitors (Table 9.1). Co^{2+} competitively inhibits Mg^{2+} and Ni^{2+} uptake but is not transported by either system. MgtA and MgtB differ in their sensitivity to inhibition by the cations Ca^{2+} and Zn^{2+}. Ca^{2+} inhibits Mg^{2+} uptake by MgtA with a K_i of about 300 μM.[57] Uptake via the MgtB system is not inhibited by Ca^{2+} concentrations exceeding 30 mM, four orders of magnitude greater than the K_m for Mg^{2+} influx. A similar result is obtained with Zn^{2+}. Mg^{2+} uptake via MgtA is inhibited by Zn^{2+}, while influx mediated by MgtB is not affected. The cation inhibition profiles for the MgtA and MgtB systems differ

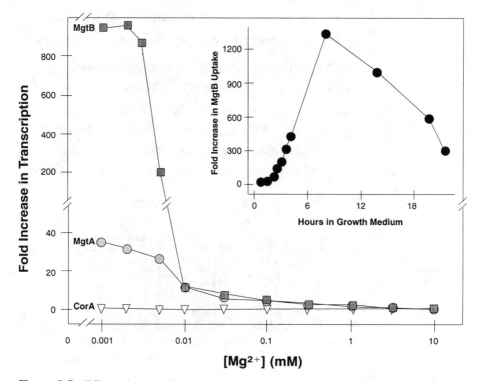

Figure 9.7 Effect of extracellular Mg^{2+} on transcription and transport. Transcriptional activity of the genes encoding the three Mg^{2+} transport systems was determined using *lacZ* operon fusions. A similar experiment (inset) measuring the rate of Ni uptake in an MgtB-only strain was performed, indicating that this dramatic increase in transcription can be translated into functional transport. At low Mg^{2+} concentrations MgtB transcription increases 1000-fold, while a more modest 40-fold increase is observed with MgtA. Transcription of the corA allele does not respond significantly to exogenous levels of Mg^{2+}.

greatly from those for the CorA system (Table 9.2). Together with their inability to transport Co^{2+}, this may suggest a difference in the structure of ligand binding site for CorA versus the Mgt proteins.

In minimal growth medium, the MgtA system transports $^{28}Mg^{2+}$ with a K_m of about 12 μM at 20°C and a V_{max} of about 24 pmol/min/10^8 cells. Under identical assay conditions, uptake via MgtB could not be detected. Transport activity was only observed as the reaction temperature was increased to 37°C. Thus the MgtB system is far more sensitive to temperature than MgtA and CorA. The MgtB system mediates the influx of $^{28}Mg^{2+}$ with a K_m of about 6 μM at 37°C and a V_{max} of about 75 pmol/min/10^8 cells at 10 μM extracellular Mg^{2+} (Table 9.1). In order to compare the kinetic properties of the two systems, Mg^{2+} uptake via MgtA was assayed at 37°C. Under these conditions the MgtA system influxes $^{28}Mg^{2+}$ with a K_m of about 30 μM and a V_{max} of about 150 pmol/min/10^8 cells. The data indicate that under nonrepressing conditions, the influx capacity of MgtB and MgtA are roughly equivalent. The additional transcriptional activity of MgtB at extremely low Mg^{2+} concentration indicates that it probably becomes the dominant Mg^{2+} transporter at low extracellular Mg^{2+} concentrations.

9.4.3.2.2 The MgtA Locus

The genetic locus encoding the *MgtA* transport system has been cloned[47] and sequenced (Tao et al., unpublished data). Recombinant plasmids encoding the *MgtA* locus were able to restore growth to a transport-deficient strain without supplemental Mg^{2+}. Sequence analysis of the *MgtA* locus identified a large open reading frame encoding a polypeptide with a deduced molecular mass of 95–99 kDa. Expression studies demonstrated a plasmid-encoded protein with a gel molecular weight of 91 kDa that was identified as the transport protein gene by complementation analysis.[49] The predicted amino acid sequence of the MgtA protein is approximately 50% identical to MgtB and, when conserved substitutions are considered, 75% similar. A second open reading frame situated 5' to the *MgtA* structural gene has also been identified. This open reading frame is transcribed in the opposite direction with respect to MgtB and encodes a protein with a gel molecular mass of 37 kDa. Partial sequence data suggest that the 37 kDa protein is highly homologous to the family of operon repressors related to the lactose and maltose operon repressor. No function has yet been assigned to this protein, although a role in the transcriptional modulation of *MgtA* and/or *MgtB* might be hypothesized.

9.4.3.2.3 The MgtB Locus

Sequence and protein expression data indicate that the *MgtB* locus consists of two open reading frames that have been designated *MgtC* and *MgtB*. The *MgtB* allele encodes a polypeptide with a gel molecular mass of 101 kDa, and the *MgtC* allele encodes a 22.5 kDa protein.[57] Both of these proteins are integral membrane proteins by biochemical and molecular criteria. The two genes are transcribed in a clockwise direction with respect to the chromosomal map and most likely comprise an operon with transcription initiating from the upstream *MgtC* locus.[51] Preliminary experiments have failed to identify a functional *MgtB* promoter in the region between the two loci (Tao et al., unpublished data) but show a powerful, Mg^{2+}-repressible promoter upstream of *MgtC*, indicating that *MgtC* and *MgtB* are likely transcribed as an operon. The 101 kDa *MgtB* gene product has been identified as the primary structural gene for the MgtB transport system by complementation analysis. It has not been determined whether the *MgtC* gene product is also required for transport function.

The *MgtB* open reading frame encodes a polypeptide 908 amino acid residues in length with a predicted molecular mass of 102 kDa, consistent with expression data.[49] Comparisons of the nucleotide and predicted amino acid sequences with the Genbank and EMBL databases indicate significant degree of homology with the ion-motive P-type ATPase gene family.[58–60] The P-type ATPase proteins are a group of cation transport systems whose members transport all of the major biologically relevant cations, Mg^{2+}, Ca^{2+}, Na^+, K^+, and H^+. These proteins utilize a transport mechanism that involves a phosphorylation–dephosphorylation cycle at a conserved aspartyl residue during each transport cycle.

While MgtB has been identified as a P-type ATPase by sequence homology, it surprisingly has little homology to the other prokaryotic proteins of this class.[61–63] The most significant prokaryotic similarity lies with the KdpB subunit of the Kdp K$^+$ influx system from *E. coli*. This protein is considerably shorter than MgtB, 682 residues as opposed to 908 residues for MgtB. The total similarity between these proteins is 30%, with some regions possessing as much as 65% homology. The most significant homology for MgtB lies with the sarco(endo)plasmic reticular Ca^{2+} ATPase (SERCA) proteins of yeast and of mammalian skeletal muscle.[64–66] The length of the SERCA proteins are 950–1000 amino acid residues, similar to the 908 residues of MgtB, and the MgtB and SERCA proteins are 50% similar, with some regions possessing similarities greater than 80% (Fig. 9.8).

A.   ATPase Phosphorylation Site

KdpB   PTTIGGLLSASAVAGMSRMLGAN-VIATSGRAVEAAGDVDVLLLDKTGTIT   313
         | :   :: :| :| :   : :: || |:: | :| |-| |||||:|
NC   PVTLP-AVTTT-MAVGAAYLAKK-AIVQKLSAIESLAGVEILCSDKTGTLT   383
         | || : :: :| || :::: || :| ||:   :::::||:||||||||
MgtB   PEMLP-MIVSSNLAKGAIAMSRRKVIVKRLNAIQNFGAMDVLCTDKTGTLT   385
         || || :||: || | : |::| ||:|| ::: |:::|:|:||||||||
PMR1   PEGLP-IIVTVTLALGVLRMAKRKAIVRRLPSVETLGSVNVICSDKTGTLT   377
         || || :::: || |: |::: ||: | ::: | |:|:||||||||
CaST   PEGLP-AVITTCLALGTRRMAKKNAIVRSLPSVETLGCTSVICSDKTGTLT   357
                                                     *

B.   "Hinge" Region

KdpB   TPEAKLALIRQYQAEGRLVAMTGDGTNDAPALAQADVAVAMNS-GTQAAKEAGNMVDLD   552
         || | :: | | |: ||| ||||||| :|||::::::: ::: ||::::: |:
NC   FPQHKYNVVEILQQRGYLVAMTGDGVNDAPSLKKADTGIAVEG-SSDAARSAADIVFLA   668
         | | ::: || | |: |||:|||||:|: || ||:|:: ::| : ::||: |
MgtB   TPLQKTRILQALQKNGHTVGFLGDGINDAPALRDADVGISVDS-AADIRKESSDIILLE   684
         || || :||: || |" || |"|||||| "|"|"| "":| ||"||""|:
PMR1   TPEHKLNIVRALRKRGDVVAMTGDGVNDAPALKLSDIGVSMGRIGTDVAKEASDMVLTD   715
         | |::|:: || : ||||:|||||||: |::|::: | :: : | :|:::| :
CaST   EPSHKSKIVEFLQSFDEITAMTGDGVNDAPALKKAEIGIAMGS-GTAVAKTASEMVLAD   736

Figure 9.8 Homology of MgtB with other P-type ATPases. Alignments of the amino acid sequences for the representative P-type ATPases are shown in two highly conserved regions. Region A is the sequence flanking the conserved aspartyl residue that is phosphorylated during the reaction cycle. The phosphorylated aspartate residue is indicated by an asterisk. Region B is the putative "hinge" sequence that may participate in a conformational change of the cytoplasmic domain during the reaction cycle. MgtB residues that are identical to those in other sequences are indicated by [|], while residue that represent conserved substitutions are indicated by [:]. Groups of similar amino acids were defined as (G,A,S,T), (L,I,V,M), (H,R,K), (D,E,N,Q), and (Y,F,W).[58] The numbers to the right of each line indicate the residue number of the last amino acid on that line in the respective protein. KdpB is the β subunit of the *E. coli* Kdp K$^+$-ATPase.[61] NC is the plasma membrane H$^+$-ATPase of *Neurospora crassa*.[82] PMR1 is a Ca^{2+}-ATPase of *Saccharomyces cerevisiae*.[66] CaST is the sarcoplasmic reticulum Ca^{2+}-ATPase from rabbit slow twitch skeletal muscle.[69] Alignments were performed as described by Snavely et al.[58]

9.4.3.2.4 Membrane Topology of MgtB

Hydropathy profiles for MgtB and the SERCA Ca^{2+} transport proteins, as well as the other eukaryotic P-type ATPases, are remarkably similar, suggesting they that share a common membrane topology.[67] Biochemical studies on several P-type ATPase proteins indicate that their N and C termini are localized to the cytosol, thus requiring an even number of membrane spanning segments. The exact number cannot be unambiguously determined solely from the hydropathy plots and has been the source of considerable speculation. Serrano has proposed a model with eight transmembrane segments while MacLennan and co-workers predict ten (Fig. 9.9).[68,69] Other models ranging from 7 to 12 segments have been proposed. Our laboratory recently defined the topology of the MgtB protein using a combination of protein fusions and paired epitope insertions.[67] The data indicate a model for MgtB topology that contains ten transmembrane helices with a cytoplasmic disposition of both the N and C termini (Fig. 9.10). This model is most similar to the ten-segment model initially proposed by Brandl et al. for the Ca^{2+} ATPase from sarcoplasmic reticulum and provides a model for the elucidation of the details of P-type ATPase structure.[69]

9.4.3.2.5 The Role of MgtC

While most of the known P-type ATPase systems consist of a single α subunit, the Na^+,K^+- and H^+,K^+-ATPases have an additional β subunit. These proteins are much smaller than the α moiety and appear to play a role in the proper membrane insertion of the α subunit.[70-72] While initial genetic and biochemical studies characterized MgtC as an integral membrane protein, its expression does not appear to be an absolute requirement for transport. Recent studies have demonstrated that MgtB correctly inserts into the cytoplasmic membrane in the absence of MgtC (T. Tao and M. E. Maguire, unpublished data). Thus MgtC does not appear to function like the known β subunits. Furthermore, the sequence of MgtC is not homologous to any of the previously described β subunits, although the members of this group do not possess a major degree of sequence homology with each other.[73-75] Another possibility for the function of MgtC is that of a binding protein. Topological predictions,[76] electron microscopic,[77] and two-dimensional X-ray crystallographic[78] data indicate that P-type ATPases have a general structure that does not protrude into the exoplasmic space. This structure, coupled with the fact that the only P-type ATPases that have β subunits are influx systems, suggest that a β subunit could facilitate the binding of the *extracellular* cation and subsequent delivery of cation to the membrane-embedded ATPase subunit.

9.5 Conclusion

Mg^{2+} transport systems have been described in a variety of prokaryotic and eukaryotic organisms. Studies with squid axon, barnacle muscle, and erythrocytes have identified a widespread Mg^{2+}/Na^+ antiport efflux system that appears to be a dominant mechanism for the maintenance of Mg^{2+} homeostasis within the cell. An

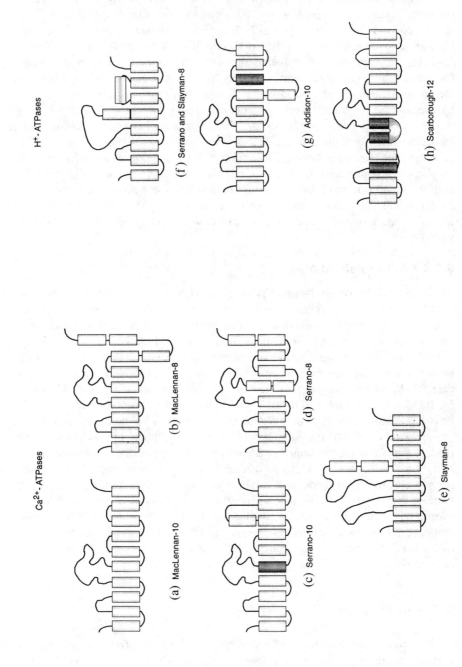

Ca²⁺ - ATPases

(a) MacLennan-10

(b) MacLennan-8

(c) Serrano-10

(d) Serrano-8

(e) Slayman-8

H⁺ - ATPases

(f) Serrano and Slayman-8

(g) Addison-10

(h) Scarborough-12

additional Mg^{2+} influx system that does not exhibit a Na^+ dependency has been described in murine lymphoma cells. These studies have been limited to gross kinetic characterization of transport because of the technical difficulties associated measuring Mg^{2+} and the inherent genetic limitations of the eukaryotic organisms. By far the most detailed studies of Mg^{2+} transport processes have been in the prokaryotic organisms *Salmonella typhimurium* and *Escherichia coli*. These studies have resulted in the first characterization of a Mg^{2+} transport system at the molecular level.

The MgtA and MgtB systems of *Salmonella typhimurium* clearly belong to the well-described family of P-type ATPase transport proteins. Nonetheless they are significantly different from the H^+, K^+, Na^+, and Ca^{2+} P-type ATPase proteins of this family in two respects. First, the amino acid sequences of the MgtA and MgtB proteins are more similar to eukaryotic transport proteins than to their prokaryotic homologues. This observation suggests that MgtA and MgtB form the first members of a subfamily of prokaryotic P-type ATPases that may be the evolutionary precursors of divalent-cation-transporting P-type ATPases in eukaryotes. Second, it is unclear as to why *S. typhimurium* would use ATP to mediate Mg^{2+} influx since the electrochemical gradient for Mg^{2+} is directed inward and would be more than sufficient to facilitate Mg^{2+} influx. It has not been determined whether this additional energy is required to couple Mg^{2+} transport to that of another ion, or whether the ATPase activity is necessary under particular growth conditions. Recent data have indicated that transcription of the *MgtB* allele is significantly increased as *S. typhimurium* is phagocytized into vacuoles of mammalian epithelial cells,[79] suggesting the hypothesis that this transporter has a role in pathogenesis.

The CorA Mg^{2+} transport systems of *S. typhimurium* and *E. coli* appear to define a novel class of transport proteins. Their sequence and membrane topology do not resemble that of any previously described transport protein. CorA is constitutively expressed and serves as the primary means of accumulating Mg^{2+} under laboratory growth conditions. At such Mg^{2+} concentrations, neither the MgtA nor MgtB systems are expressed at significant levels, while the CorA system functions at maximum capacity. Moreover, the V_{max} and K_m values for Mg^{2+} transport indicate

---

Figure 9.9 Predicted models for the membrane topology of various P-type ATPases. Topology models are shown for several types of P-type ATPase proteins. The models are compared to the ten membrane segment model for the sarcoplasmic reticular Ca^{2+} P-type ATPase of rabbit slow twitch skeletal muscle.[69] Additional models are shown with the same membrane spanning segments in identical order as open boxes either within or outside the cytoplasmic membrane as predicted by that model. Additional membrane segments predicted by some models are shown as darker hatched boxes. The models are attributed to the laboratory of origin. (a) MacLennan ten-segment model for Ca^{2+}-ATPase.[64,82] (b) MacLennan eight-segment model for Ca^{2+}-ATPase.[64,82] (c) Serrano ten-segment model for Ca^{2+}-ATPase.[83] (d) Serrano eight-segment model for Ca^{2+} ATPase.[68] (e) Slayman eight-segment model for Ca^{2+}-ATPase.[84] (f) Serrano and Slayman models for H^+-ATPase from *S. cerevisiae* and *N. crassa*.[68,84,85] (g) Addison ten-segment model for *S. cerevisiae* H^+-ATPase.[86] (h) Scarborough 12-segment model for *N. crassa* H^+-ATPase.[85,87]

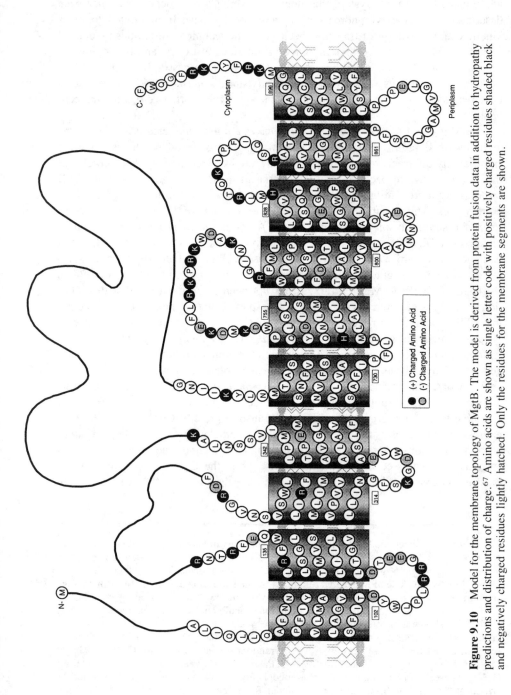

Figure 9.10 Model for the membrane topology of MgtB. The model is derived from protein fusion data in addition to hydropathy predictions and distribution of charge.[67] Amino acids are shown as single letter code with positively charged residues shaded black and negatively charged residues lightly hatched. Only the residues for the membrane segments are shown.

that the CorA system alone is more than capable of satisfying the cellular requirement for Mg^{2+}. If a minimal requirement for Mg^{2+} accumulation is a doubling of intracellular Mg^{2+} concentration within a single cell cycle (~60 minutes in the laboratory), the transport capacity of the CorA system is more than sufficient since it could double the cell's Mg^{2+} content in less than 1 minute. This excess capacity and the individual physiological roles of these three distinct transport Mg^{2+} systems thus remain a viable area for future experimentation.

Previous studies have described CorA-like systems in bacteria as diverse as the photosynthetic bacterium *Rhodobacter capsulatus* and the gram-positive bacterium *Bacillus subtilis*. Our laboratory has recently cloned loci-encoding transport systems having a CorA-like phenotype from *Providencia stuartii, Enterococcus cass,* and *Bacillus firmus* OF4. Thus while CorA represents a new class of transport protein, it may be ubiquitous within the prokaryotic kingdom. Studies directed towards identifying and characterizing CorA homologues in eukaryotic organisms are an obvious next step.

References

1. Reinhart, R. A., *Arch. Intern. Med.* **1988,** *148,* 2415–20.

2. Rotevatn, S.; Sarheim, H.; Murphy, E., *Acta Physiol. Scand.* **1991,** *142* (Suppl. 599), 125–33.

3. Rotevatn, S.; Murphy, E.; Levy, L. A.; Raju, B.; Lieberman, M.; London, R. E., *Am. J. Physiol.* **1989,** *257,* C141–46.

4. London, R. E., *Ann. Rev. Physiol.* **1991,** *53,* 241–58.

5. Murphy, E.; Freudenrich, C. C.; Lieberman, M., *Ann. Rev. Physiol.* **1991,** *53,* 273–87.

6. White, R. E.; Hartzell, H. C., *Biochem. Pharmacol.* **1989,** *38,* 859–67.

7. Matsuda, H., *Ann. Rev. Physiol.* **1991,** *53,* 289–98.

8. Walker, G. M., *Magnesium* **1986,** *5,* 9–23.

9. Maguire, M. E., *Ann. NY Acad. Sci* **1988,** *551,* 201–17.

10. Erdos, J. J.; Maguire, M. E., *J. Physiol.* **1983,** *337,* 351–71.

11. Grubbs, R. D., *Am. J. Physiol. Cell Physiol.* **1991,** *260,* C1158–64.

12. Romani, A.; Scarpa, A., *Arch. Biochem. Biophys.* **1992,** *298,* 1–12.

13. Dai, L. J.; Quamme, G. A., *Am. J. Physiol.* **1992,** *262,* F1100–4.

14. Walker, G. M.; Duffus, J. H., *Magnesium* **1983,** *2,* 1–16.

15. Seelig, M., *Am. J. Cardiol.* **1989,** *63,* 4G–21G.

16. Kass, E. H., *Rev. Infect. Dis.* **1989,** *11* (suppl.), S167–75.

17. Maguire, M. E., *TIBS* **1984,** *5,* 73–77.

18. Maguire, M. E., *Metal Ions Biol.* **1990,** *26,* 135–53.

19. Grubbs, R. D., *Metal Ions Biol.* **1990,** *26,* 177–92.

20. Gunther, T.; Vormann, J., *Biochem. Biophys. Res. Comm.* **1985,** *130,* 540–45.

21. Feray, J. C.; Garay, R., *Biochem. Biophys. Acta.* **1986,** *856,* 76–84.

22. Ludi, H.; Schatzmann, H. J., *J. Physiol. (Lond.)* **1987,** *390,* 367–82.

23. Gunther, T.; Ebel, H., *Metal Ions Biol.* **1990,** 26, 215–26.

24. Baker, P. F.; Crawford, A. C., *J. Physiol.* **1972,** *227,* 855–74.

25. De Weer, P., *J. Gen. Physiol.* **1976,** *68,* 159–78.

26. DiPolo, R.; Beauge, L., *Biochim. Biophys. Acta* **1988,** *946,* 424–28.

27. Feray, J. C.; Garay, R., *Arch. Pharmacol.* **1988,** *338,* 332–37.

28. Mullins, L. J.; Brinley, F. J.; Spangler, S. G.; Abercrombie, R. F., *J. Gen. Physiol.*, **1977,** *69,* 389–400.

29. Flatman, P. W.; Smith, L. M., *J. Physiol.* **1990,** *431,* 11–25.

30. Flatman, P. W.; Smith, L. M., *J. Physiol.* **1991,** *443,* 217–30.

31. Maguire, M. E.; Erdos, J. J., *J. Biol. Chem.* **1978,** *253,* 6633–36.

32. Maguire, M. E.; Erdos, J. J., *J. Biol. Chem.* **1980,** *255,* 1030–35.

33. Erdos, J. J.; Maguire, M. E., *Mol. Pharmacol.* **1980,** *18,* 379–83.

34. Quamme, G. A.; Dai, L. J., *Am. J. Physiol.* **1990,** *259,* C521–25.

35. Henderson, P.J.F.; Maiden, M.C.J., *Phil. Trans. R. Soc. London (Biol.)* **1990,** *326,* 391–410.

36. Epstein, W.; Walderhaug, M. O.; Polarek, J. W.; Hesse, J. E.; Dorus, E.; Daniel, J. M., *Phil. Trans. R. Soc. London (Biol.)* **1990,** *326,* 479–87.

37. Blight, M. A.; Holland, I. B., *Mol. Microbiol.* **1990,** *4,* 873–80.

38. Ames, G. F.-L.; Mimura, C. S.; Shyamala, V., *FEMS Microbiol. Microbiol. Rev.* **1990,** *75,* 429–46.

39. Maloney, P. C., *FEMS Microbiol. Rev.* **1990,** *87,* 91–102.

40. Silver, S., *Proc. Nat. Acad. Sci. USA* **1969,** *62,* 764–71.

41. Silver, S.; Clark, D., *J. Biol. Chem.* **1971,** *246,* 569–71.

42. Lusk, J. E.; Kennedy, E. P., *J. Bacteriol.* **1969,** *144,* 1653–55.

43. Nelson, D. L.; Kennedy, E. P., *J. Biol. Chem.* **1971,** *246,* 3042–49.

44. Nelson, D. L.; Kennedy, E. P., *Proc. Nat. Acad. Sci. USA* **1972,** *69,* 1091–93.

45. Park, M. H.; Wong, B. B.; Lusk, J. E., *J. Bacteriol.* **1976,** *126,* 1096–103.

46. Hmiel, S. P.; Snavely, M. D.; Miller, C. G.; Maguire, M. E., *J. Bacteriol.* **1986,** *168,* 1444–50.

47. Hmiel, S. P.; Snavely, M. D.; Florer, J. B.; Maguire, M. E.; Miller, C. G., *J. Bacteriol.* **1989,** *171,* 4742–51.

48. Silver, S.; Lusk, J. E., In *Ion Transport in Prokaryotes* (Eds. Rosen, B. P.; Silver, S.), 1987, pp. 165–80, Academic Press, San Diego, CA.

49. Snavely, M. D.; Florer, J. B.; Miller, C. G.; Maguire, M. E., *J. Bacteriol.* **1989,** *171,* 4761–66.

50. Gibson, M. M.; Bagga, D. A.; Miller, C. G.; Maguire, M. E., *Mol. Microbiol.* **1991,** *5,* 2753–62.

51. Snavely, M. D.; Florer, J. B.; Miller, C. G.; Maguire, M. E., *J. Bacteriol.* **1989,** *171,* 4752–60.

52. Smith, R. L.; Banks, J.; Snavely, M.; Maguire, M. E., *J. Biol. Chem.* **1993**, *268*, 14071–80.

53. Manoil, C.; Boyd, D.; Beckwith, J., *TIG* **1988**, *4*, 223–26.

54. Manoil, C.; Boyd, D.; Beckwith, J., *Proc. Natl. Acad. Sci. USA* **1987**, *84*, 8525–29.

55. Manoil, C., *J. Bacteriol.* **1990**, *172*, 1035–42.

56. Broome-Smith, J. K.; Tadayyon, M.; Zhang, Y., *Mol. Microbiol.* **1990**, *4*, 1637–44.

57. Snavely, M. D.; Gravina, S. A.; Cheung, T. T.; Miller, C. G.; Maguire, M. E., *J. Biol. Chem.* **1991a**, *266*, 824–29.

58. Snavely, M. D.; Miller, C. G.; Maguire, M. E., *J. Biol. Chem.* **1991b**, *266*, 815–23.

59. Pederson, P. L.; Carafoli, E., *TIBS* **1987a**, *12*, 146–50.

60. Pederson, P. L.; Carafoli, E., *TIBS* **1987b**, *12*, 186–89.

61. Heese, J. E.; Wieczorek, L.; Altendorf, K.; Reicin, A. S.; Dorus, E.; Epstein, W., *Proc. Natl. Acad. Sci. USA* **1984**, *81*, 4746–50.

62. Walderhaug, M. O.; Litwack, E. D.; and Epstein, W., *J. Bacteriol.* **1989**, *171*, 1192–95.

63. Epstein, W., In *The Bacteria, Vol. XII: Bacterial Energetics* (Ed. Krulwich, T. A.), 1990, pp. 87–110, Academic Press, New York.

64. MacLennan, D. H.; Brandl, C. J.; Korczak, B.; Green, N. M., *Nature* **1985**, *316*, 696–700.

65. Inisi, G.; Sumbilla, C.; Kirtley, M. E., *Physiol. Rev.* **1990**, *70*, 749–60.

66. Rudolph, H. K.; Antebi, A.; Fink, G. R.; Buckley, C. M.; Dorman, T. E.; Le Vitre, J.; Davidow, L. S.; Mao, J.; Moir, D. T., *Cell* **1989**, *58*, 133–45.

67. Smith, D. L.; Tao, T.; Maguire, M. E., *J. Biol. Chem.* **1993**, *268*, 22469–79.

68. Serrano, R., *Biochim. Biophys. Acta* **1988**, *947*, 1–28.

69. Brandl, C. J.; Green, N. M.; Korczak, B.; MacLennan, D. H., *Cell* **1986**, *44*, 597–607.

70. McDonnough, A. A.; Geering, K.; Farley, R. A., *FASEB J.* **1990**, *4*, 1598–603.

71. Geering, K., *J. Memb. Biol.* **1990**, *115*, 109–21.

72. Geering, K., *FEBS Lett.* **1991**, *285*, 189–93.

73. Shull, G. E., *J. Biol. Chem.* **1990**, *265*, 12123–26.

74. Canfield, V. A.; Okamoto, C. T.; Chow, D.; Dorfman, J.; Gros, P.; Forte, J. G.; Levenson, R., *J. Biol. Chem.* **1990**, *265*, 19878–84.

75. Shyjan, A. W.; Canfield, V. A.; Levenson, R., *Genomics* **1991**, *11*, 435–43.

76. Stokes, D. L., *Curr. Opin. Struct. Biol.* **1991**, *1*, 555–61.

77. Stokes, D. L.; Green, N. M., *J. Mol. Biol.* **1990**, *213*, 529–38.

78. Blasie, J. K.; Pascolini, D.; Asturias, F.; Hebbette, L. G.; Pierce, D.; Scarpa, A., *Biophys. J.* **1990**, *58*, 687–93.

79. Portillo, F. G.; Foster, J. W.; Maguire, M. E.; Finlay, B. B., *Mol. Microbiol.* **1992**, *6*, 3289–97.

80. Grubbs, R. D.; Snavely, M. D.; Hmiel, S. P.; Maguire, M. E., *Methods Enzymol.* **1989**, *173*, 546–63.

81. Hille, B., *Ionic Channels of Excitable Membranes*, Sinauer, Sunderland, MA, 1992.

82. Clark, D. M.; Loo, T. W.; Inesi, G.; MacLennan, D. H., *Nature* **1989**, *339*, 476–78.

83. Serrano, R.; Kielland-Brandt, C.; Fink, G. R., *Nature* **1986,** *319,* 689–93.

84. Nakamoto, R. K.; Rao, R.; Slayman, C. W., *Ann. N.Y. Acad. Sci.* **1989,** *574,* 165–79.

85. Subrahmanyeswara Rao, U.; Hennessey, J. P., Jr.; Scarborough, G. A., *J. Biol. Chem.* **1991,** *266,* 14740–46.

86. Addison, R., *J. Biol. Chem.* **1986,** *261,* 14896–901.

87. Scarborough, G. A.; Hennessey, J. P., Jr., *J. Biol. Chem.* **1990,** *265,* 16145–49.

CHAPTER
10

Regulation of Cytosolic Magnesium Ion in the Heart

A. Romani and A. Scarpa

10.1 Introduction

The purpose of this review is to summarize recent findings on the cellular regulation of magnesium in the heart and in its subcellular organelles. The concentration of total Mg^{2+} in heart muscle cells is high, with values between 17 and 11 mM having been reported in the literature (Table 10.1).[1–11] Similar to other tissues in eukaryotic cells, the majority of Mg^{2+} in the heart is bound to ATP,[12] other binding proteins,[13] solutes, or internalized within intracellular structures.[12,14] The reported values of free Mg^{2+} in the heart under physiological conditions have a broader range because the techniques used are either indirect[2,5–7] or the values obtained are difficult to quantify.[1,3,5,7,10,11]

Table 10.1 presents the data that have been recently published by various laboratories using the equilibria of Mg^{2+}-sensitive cellular enzymes,[8] the spectral shift of ATP by ^{31}P NMR,[3] the use of fluorescent indicators,[11] the use of selective Mg^{2+} electrodes and intracellular recordings,[5–7] or [28]Mg^{2+} distribution.[1,9] Whereas the absolute value for cytosolic Mg^{2+} lies within a broad range (0.4 to 3.5 mM), there is a clear consensus that, whatever its initial value may be, cytosolic free Mg^{2+} does not change within the cell or changes minimally even under drastic perturbation of metabolic conditions.[11–15]

It was not until recently that a large redistribution of total Mg^{2+} in either direction in the heart or in myocytes was identified and characterized.[10,16,17] Under conditions that may resemble physiological conditions in vivo, such as hormonal stimulation, it was found that 5–10% of the total magnesium leaves or enters the cell within a few minutes.[10,16,17] The following two facts lead to the conclusion that

Table 10.1 TOTAL AND FREE MAGNESIUM IN MAMMALIAN
CARDIAC CELLS

Total $[Mg^{2+}]$ (mM)	Free $[Mg^{2+}]$ (mM)	Method	Ref.
17	1.0	^{28}Mg	1
	1.5–3.6	Mg flux	2
	0.8	^{31}P-NMR	3
	2.5	^{31}P-NMR[a]	4
	3.0–3.5	Mg^{2+} electrode	5
	0.4	Mg^{2+} electrode	6
	2.4	Mg^{2+} electrode	7
	0.44		
17	0.44	^{13}C-NMR citrate/isocitrate ratios	8
11–12		^{28}Mg redistribution	9
11–12	0.5–0.7	Atomic absorbance	10
	ca. 0.5	Mg^{2+} fluorescent dye	11

[a] It has been demonstrated by Gupta et al. and also by other groups that a wrong K_d has been used to estimate these data.[3]

under these conditions Mg^{2+} must redistribute from or to intracellular organelles. First, this amount is equal to or larger than the entire cytosolic free Mg^{2+}, and second, under these conditions little or no change in free Mg^{2+} is observed. Experiments recently published and reviewed here[9,10,16,17] are consistent with the view that cytosolic *free* Mg^{2+} does not change over a broad range of physiological conditions. In contrast, large fluxes of Mg^{2+} across the plasma membrane in either direction are observable, resulting in a change of total, but not free, Mg^{2+}.

10.2 Mg^{2+} Efflux from Perfused Rat Hearts

Figure 10.1(a) shows the result of an experiment where Mg^{2+} released from a perfused rat heart was measured. To permit measurements by atomic absorbance (AA) spectrophotometry, 10 min before the zero time of the trace, the perfusion medium was switched to one containing no added Mg^{2+}. Each bar of the trace represents the amount of Mg^{2+} (released plus contaminant) collected every 20 s in the perfusate, and measured by AA spectrophotometry. Mg^{2+} content in the perfusate was 7.5 μM and decreased to 3–4 μM over a 30 min perfusion period. By contrast, Figure 10.1(b) shows that the addition of norepinephrine (NE) after 5 min results in a major Mg^{2+} efflux from the heart. This efflux (3–4-fold more Mg^{2+} in the perfusate) continued for 5–10 min after NE addition. In different hearts, it was found that the NE-stimulated Mg^{2+} efflux was dose dependent; the smallest effect observable at 0.04 μM and the largest [approximately twofold that of Fig. 10.1(b)] at 1 μM NE. By challenging the heart with short stimulation of NE, it was found that each successive stimulation resulted in a progressively smaller efflux.[17] While we originally concluded that these results could be accounted for by either a down regulation of adrenergic receptors or a depletion of Mg^{2+} from intracellular

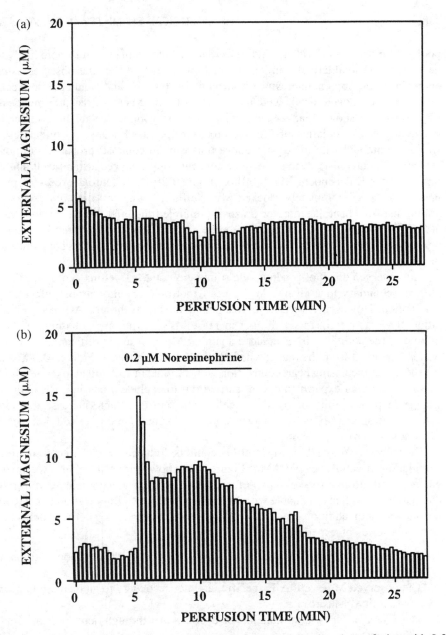

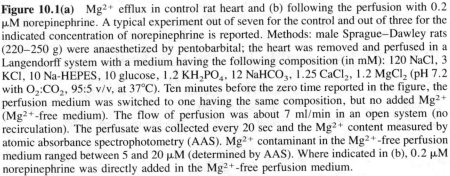

Figure 10.1(a) Mg^{2+} efflux in control rat heart and (b) following the perfusion with 0.2 μM norepinephrine. A typical experiment out of seven for the control and out of three for the indicated concentration of norepinephrine is reported. Methods: male Sprague–Dawley rats (220–250 g) were anaesthetized by pentobarbital; the heart was removed and perfused in a Langendorff system with a medium having the following composition (in mM): 120 NaCl, 3 KCl, 10 Na-HEPES, 10 glucose, 1.2 KH_2PO_4, 12 $NaHCO_3$, 1.25 $CaCl_2$, 1.2 $MgCl_2$ (pH 7.2 with $O_2{:}CO_2$, 95:5 v/v, at 37°C). Ten minutes before the zero time reported in the figure, the perfusion medium was switched to one having the same composition, but no added Mg^{2+} (Mg^{2+}-free medium). The flow of perfusion was about 7 ml/min in an open system (no recirculation). The perfusate was collected every 20 sec and the Mg^{2+} content measured by atomic absorbance spectrophotometry (AAS). Mg^{2+} contaminant in the Mg^{2+}-free perfusion medium ranged between 5 and 20 μM (determined by AAS). Where indicated in (b), 0.2 μM norepinephrine was directly added in the Mg^{2+}-free perfusion medium.

pool(s),[17] more recent evidence indicates that the latter explanation is more likely to be accurate. A qualitatively similar stimulation of Mg^{2+} efflux was observed using epinephrine or isoprenaline. Since heart contains both α- and β-adrenergic receptors,[18] α and β antagonists[19] were used to identify the type(s) of receptor involved. Figure 10.2(a) shows that α_2 antagonists such as yohimbine had little effect in preventing NE stimulation of Mg^{2+} efflux from perfused hearts. A similar result was also obtained with the α_1 antagonist prazosin. In contrast, propanolol (a nonselective β antagonist), even at the same concentration, induced a significant inhibition of the NE-dependent Mg^{2+} efflux [Fig. 10.2(b)]. A tenfold excess of propanolol over NE completely blocked Mg^{2+} efflux,[17] whereas a tenfold excess of yohimbine (or phentolananine or prazosi) inhibited the effect of NE by less than 50%.[20] These effects, which could be even more clearly demonstrated in isolated myocytes,[17] lead to the conclusion that activation of β-adrenergic receptors results in a major release of Mg^{2+} from perfused hearts.

The observed effect of β-adrenergic stimulation on Mg^{2+} efflux could, in principle, be secondary to a change in the heart inotropic or chronotropic effect of β stimulation. However, it was shown that pacing the hearts through A-V node stimulation at variable frequencies did not increase Mg^{2+} efflux above baseline.[10] Furthermore, the addition of NE induced a similar Mg^{2+} efflux in different hearts that was independent from the rate at which the hearts were paced.[10] Figure 10.3(a) and (b) show that increasing heart contraction with caffeine or KCl failed to cause Mg^{2+} efflux. Additional experiments were carried out to exclude a possible NE-induced change in permeability or fractional cell death. No detectable increase in lactic dehydrogenase activity nor any increase in K^+ was measured after addition of NE (data not shown).

The stimulation by NE of Mg^{2+} efflux could be induced by any direct or indirect stimulation of cellular cyclic AMP. Figure 10.4 shows the result of an experiment where the addition of forskolin to perfused hearts prompts a major release of Mg^{2+}, comparable with that obtainable with NE. A similar Mg^{2+} efflux was obtained (data not shown) upon addition of various cell permeant cAMP analogs. The conclusions from this series of experiments are as follows:

1. A large Mg^{2+} efflux can be induced in isolated perfused hearts upon addition of norepinephrine.
2. The observed Mg^{2+} efflux is specific, and not secondary to an increase in heart contractility or heart rate.
3. The effect occurs via stimulation of cyclic AMP through activation of β adrenoreceptors.

10.3 Mg^{2+} Movement in Isolated Myocytes

The effect of β-adrenergic stimulation on Mg^{2+} efflux from perfused hearts is difficult to quantify as a result of cellular heterogeneity in the heart and because of the variability of perfusion rates and other metabolic parameters of cells within the

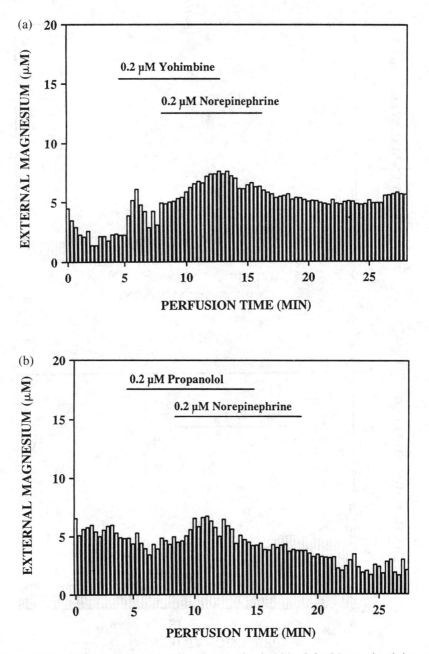

Figure 10.2 Mg^{2+} efflux in perfused rat hearts stimulated by 0.2 μM norepinephrine and pretreated with (a) 0.2 μM yohimbine or (b) 0.2 μM propanolol. One experiment typical of two for both inhibitors is reported. Methods: rat hearts were perfused and Mg^{2+} efflux was measured as reported in legend of Figure 10.1. Five minutes before the addition of norepinephrine, the α_2-antagonist yohimbine or the β-antagonist propanolol was directly added in the perfusion medium for the length of time reported in the figure.

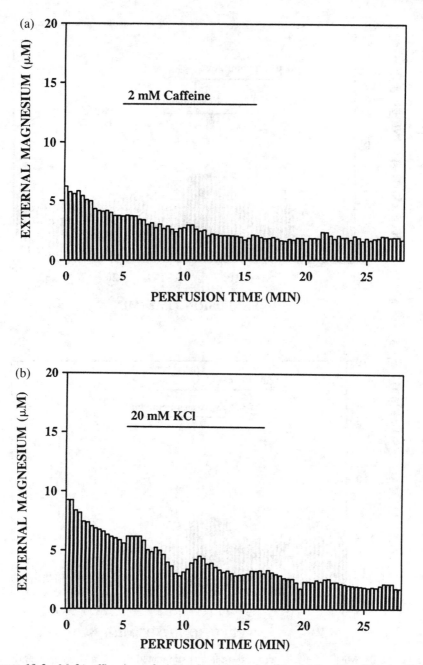

Figure 10.3 Mg^{2+} efflux in perfused rat hearts stimulated by (a) 2 mM caffeine or (b) 20 mM KCl. One experiment out of two for both caffeine and KCl is reported. Methods: rat heart perfusion and Mg^{2+} measurement in the perfusate were carried out as reported in the caption of Figure 10.1. (a) Caffeine or (b) KCl was directly added in the perfusion medium.

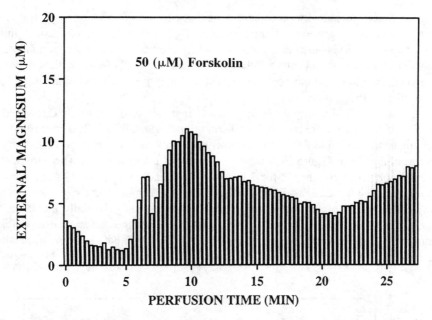

Figure 10.4 Mg^{2+} efflux in perfused rat heart stimulated by 50 μM forskolin. One typical experiment out of three is reported. Methods: rat heart was perfused and Mg^{2+} efflux was measured as reported in the caption to Figure 10.1.

same heart. This limitation can be overcome by using collagenase-dispersed rat heart myocytes. In a large number of experiments by us it was shown that the addition of NE to a suspension of myocytes stimulates a release of Mg^{2+} similar to that observable in perfused hearts.[10,17] The characteristics of this Mg^{2+} efflux can be summarized as follows:

1. Mg^{2+} efflux can be followed kinetically over a 3–5 min period following addition of NE.[17]
2. The efflux is nearly completely blocked by β-adrenergic inhibitors such as propanolol or sotalol, and only slightly inhibited by α-adrenergic inhibitors such as yohimbine, prazosin, or phenilephrine.[17,20]
3. Forskolin and permeant cAMP analogues mimic quantitatively the effect of NE.[17]
4. The amount of total Mg^{2+} released from isolated myocytes upon stimulation with NE (or permeant cyclic AMP analogues) is sizeable and corresponds to approximately 10% of the total Mg^{2+} content of the myocytes.[9,10,17]

One possible drawback of these experiments and those using perfused hearts is that Mg^{2+} was measured in the perfusate or in the supernatant by AA spectrometry. This method, although accurate and selective, cannot be carried out in the presence of physiological concentrations of Mg^{2+} in the perfusate or in the extracellular fluid. Hence, to gain the necessary sensitivity, it was necessary to perform the experiments in the presence of unphysiologically low concentrations of extracellular Mg^{2+} (5–20 μM). These concentrations, which are one-to-two orders of magnitude smaller than those found extracellularly, create an artificially large gradient of Mg^{2+} between the intra- and extracellular milieu and may have resulted in an amplification of the Mg^{2+} efflux.

The annual availability of $^{28}Mg^{2+}$ made it possible to measure NE-stimulated Mg^{2+} efflux under more physiological conditions, where the Mg^{2+} concentration in the incubation medium surrounding the myocytes closely resembles that of the extracellular fluid. Figure 10.5 shows the result of the experiment where myocytes were incubated with $^{28}Mg^{2+}$ for 3 h, a condition under which most of the isotopic equilibration between $^{28}Mg^{2+}$ and cellular Mg^{2+} is achieved. At zero time, before any addition, the amount of Mg^{2+} in three samples was essentially the same. In untreated myocytes, Mg^{2+} content was virtually unchanged after incubation for 5 min at 37°C. In contrast, treatment with NE (or forskolin; data not shown) resulted in a loss of $^{28}Mg^{2+}$ from the cell corresponding to 2 nmol/mg protein. It is notewor-

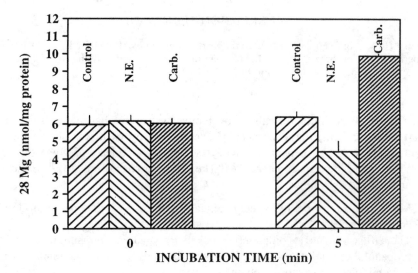

Figure 10.5 $^{28}Mg^{2+}$ redistribution in cardiac ventricular myocytes. Cardiac ventricular myocytes were isolated by collagenase digestion according to the procedure of De Young et al.,[26] and resuspended in the medium reported in the caption to Figure 10.1, in the presence of 1.2 mM $MgCl_2$ labeled with 1 $\mu Ci/ml$ $^{28}Mg^{2+}$. After 3 h, cells were washed three times with the same medium containing only "cold" $MgCl_2$ to remove excess radioisotope, and incubated therein at 37°C in the presence of $O_2:CO_2$ (95:5 v/v), at the final concentration of approximately 350 $\mu g/ml$, in the presence of 1 $\mu Ci/ml$ $^{28}Mg^{2+}$. Cells were stimulated by adding 10 μM norepinephrine (NE) or 100 μM carbachol. At the times reported in the figure an aliquot of the incubation medium was withdrawn in quadruplicate, and the cells were sedimented in microfuge tubes through an oil layer (dibutyl-phthalate:dioctyl-phthalate, 2:1).[27] Supernatant and oil layer were removed by vacuum suction and the radioactivity in the pellet was measured by liquid scintillation counting. The data are means ±S.E. of two different preparations, each one performed in duplicate.

thy that treatment with carbachol resulted in a gain of $^{28}Mg^{2+}$ in these cells. Similar Mg^{2+} uptake in the presence of carbachol was also shown in myocytes when only contaminant extracellular Mg^{2+} was present.[10,17]

10.4 Stimulation of Myocyte Mg^{2+} Uptake

We have previously shown that the addition of agents that stimulate or mimic the action of protein kinase C induce Mg^{2+} uptake across the plasma membranes.[21] This is summarized in Figure 10.6(a), which shows that phorbol 12,13-dibutyrate (PDBu), phorbol 12-myristate 13-acetate (PMA), 1-oleoyl 2-acetyl-*sn*-glycerol (OAG), and 1-stearoyl 2-arachidonoyl-*sn*-glycerol (SAG) stimulate Mg^{2+} uptake by myocytes in a way qualitatively similar to that induced by carbachol. By contrast, Figure 10.6(b) shows that inhibitors of protein kinase C, staurosporine, or 1-(5-iso-quinoline-sulfonyl)-2-methylpiperazine dihydrochloride (H7) induce Mg^{2+} release from myocytes comparable to that obtained with NE or permeant cAMP analogues.

Taken together, this and other evidence indicate a major role of protein kinase C in the stimulation of Mg^{2+} uptake in myocytes.[21] In these experiments, Mg^{2+} disappearance from the medium is being measured. The large magnitude of this uptake (5–10% of the total Mg^{2+} content of the myocytes), and the finding that no major increase of free cytosolic Mg^{2+} (unpublished results) is observable under these conditions, suggest that the Mg^{2+} accumulated across the plasma membrane is redistributed within cellular organelles. At present it is not known if the activation of protein kinase C results in the activation of a plasma membrane transporter or that of an organelle transporter, or both.

A cautionary note to these experiments is that in myocytes and in other cells there is significant cross-talk between effects mediated by protein kinase C and cAMP. Hence, some of the effects observed could be accounted for by a decrease in cAMP, which results from the agonist-derived activation of protein kinase C.[22,23]

10.5 Stimulation of Mg^{2+} Release

As discussed previously, addition of NE induced a time-dependent release of Mg^{2+} from myocytes through the activation of β-adrenergic receptors and the increase of cellular cAMP. This efflux occurs within 5 min and is quantitatively similar at concentrations of extracellular Mg^{2+} ranging from 0 to 2 mM.[10] The amount of total Mg^{2+} released within several minutes corresponds to approximately 10% of the total Mg^{2+} content of the myocytes. This amount is equal to or larger than the amount of Mg^{2+} assumed to be cytosolic free Mg^{2+} (see Table 10.1). These results, together with the finding that under NE stimulation the cytosolic free Mg^{2+} does not change appreciably,[10,15,17] strongly suggests that the majority of the Mg^{2+} released was mobilized from storage pools.

This point was extensively investigated in our laboratory in cardiac myocytes and hepatocytes.[10,24] The conclusions were that the mitochondria are the major pool

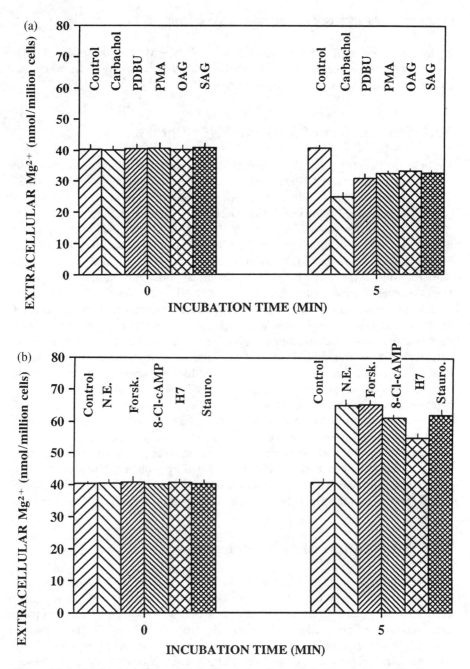

Figure 10.6 Mg^{2+} (a) influx and (b) efflux in cardiac ventricular myocytes. Cardiac ventricular myocytes were prepared according to the procedure of De Young et al.[26] Cells were incubated in the Mg^{2+}-free medium reported in the caption of Figure 10.1, at 37°C in the presence of O_2:CO_2 (95:5 v/v), at the final concentration of approximately 350 μg/ml. (a) The stimulatory effect exerted on the cells by adding 100 μM carbachol, 1 μM PDBU, 20 nM PMA, 20 μM OAG, or 20 μM SAG in the incubation medium. (b) The stimulatory effect of 10 μM norepinephrine, 50 μM forskolin, 100 μM 8-Cl-cAMP, 50 μM H7, or 50 nM staurosporine. At the times reported in the figures an aliquot of the incubation mixture was withdrawn in duplicate and the cells sedimented in microfuge tubes. Mg^{2+} content in the supernatant was measured by AAS. Data are means ±S.E. of four different preparations.

from which the measured extracellular Mg^{2+} originates upon stimulation with NE.[10,24] A representative experiment is shown in Figure 10.7. Cardiac myocytes were permeabilized with digitonin and washed once. The control trace shows that the extracellular Mg^{2+} of these permeabilized myocytes remains constant over 8 min of observation. Under these conditions, the addition of NE was without effect in mobilizing Mg^{2+} (not shown), as expected in permeabilized cells. By contrast, the addition of cAMP prompted a time-dependent Mg^{2+} release, which amounted to approximately 10 nmol/10^6 cells over 6 min. These data indicated that in permeabilized cells, cAMP releases an amount of Mg^{2+} that is quantitatively similar and has similar kinetics as that shown in intact cells upon addition of NE. Furthermore, in the absence of a functional plasma membrane, this indicates a release from a cell compartment. Figure 10.7 also shows that this Mg^{2+} release can be com-

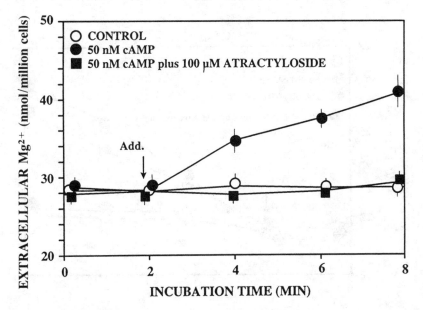

Figure 10.7 Mg^{2+} efflux from digitonin-permeabilized cardiac ventricular myocytes stimulated by 50 nM cAMP (acid form) in the absence or in the presence of 100 μM atractyloside. After collagenase digestion, cardiac ventricular myocytes were permeabilized by digitonin treatment.. The permeabilization was assessed by an ethidium bromide fluorescent test. Cells were washed three times to remove digitonin excess and incubated in a Mg^{2+}-free medium having the following composition (in mM): 100 KCl, 3 NaCl, 10 K-HEPES, 10 glucose, 10 $KHCO_3$, 1.2 KH_2PO_4, 1 $CaCl_2$, 2 EGTA, 1 ATP (plus 10 mM phosphocreatine and 10 U/ml creatine phosphokinase as regenerating system), pH 7.2 at 37°C, in the presence of O_2:CO_2 (95:5 v/v). Mg^{2+} contamination in the medium was approximately 2–10 μM (measured by AAS). Where indicated in the figure an aliquot of the incubation mixture was removed in duplicate and the cells were sedimented in microfuge tubes. The Mg^{2+} content of the supernatant was measured by AAS. 100 μM atractyloside was added to the incubation medium as a specific inhibitor of the adenine nucleotide translocase. Data are means ±S.E. of five different preparations.

pletely inhibited by actractyloside or, not shown, by bongkrekic acid.[24] As these two structurally different agents are specific inhibitors of the mitochondrial adenine nucleotide translocase, this inhibition provides strong evidence that the mitochondrion is the organelle from which Mg^{2+} is released upon cAMP addition. This was confirmed by the experiments shown in Figure 10.8, where a virtually identical experiment was carried out in the presence of isolated rat heart mitochondria instead of permeabilized cells. Again, cAMP directly stimulates Mg^{2+} release from mitochondria, in a manner similar to that observed upon the addition of ADP.[24] Again, the efflux was inhibited by actractyloside.

It is noteworthy that cAMP acts at nanomolar concentrations, whereas the effect of ADP occurs at concentrations 3–4 orders of magnitude greater. These results, as well as other preliminary experiments, have led to the conclusion that cAMP binds directly to the adenine nucleotide translocase and induces an exchange of intra-mitochondrial MgATP (instead of anionic ATP) for extramitochondrial ADP.[24,25]

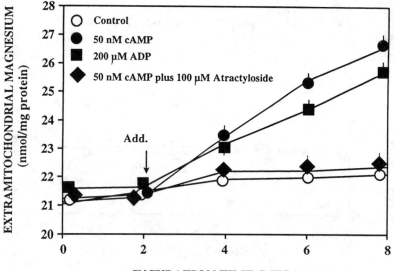

INCUBATION TIME (MIN)

Figure 10.8 Mg^{2+} efflux in isolated heart mitochondria stimulated by 50 nM cAMP or 200 μM ADP in the absence or in the presence of 100 μM atractyloside. Isolated heart mitochondria were prepared according to the procedure of Vinogradov et al.[28] Their functionality was assessed polarographically according to the procedure of Estabrook.[29] After isolation mitochondria were incubated in 200 mM sucrose, 10 mM KCl, 10 mM HEPES (pH 7.2 with Tris) at the final concentration of 350 μg/ml. At the times indicated in the figure an aliquot of the incubation mixture was withdrawn in duplicate and mitochondria were sedimented in microfuge tubes. Mg^{2+} content in the supernatant was measured by AAS. 100 μM atractyloside was added to the incubation medium as a specific inhibitor of the adenine nucleotide translocase. Data are means ±S.E. of 5 different preparations. In this as in the previous experiments protein was measured according to the procedure of Bradford.[30]

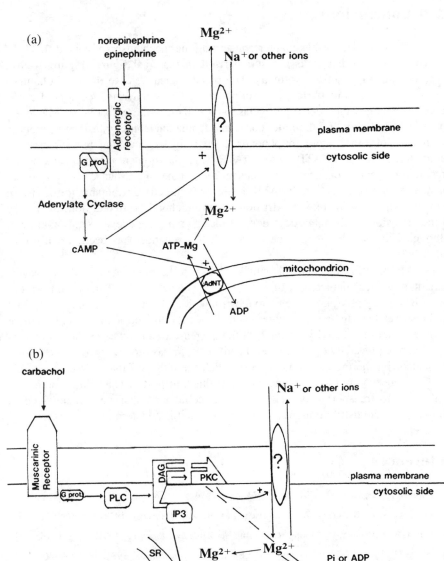

Figure 10.9 Mechanisms involved in Mg^{2+} (a) efflux and (b) influx in cardiac cells. The possible mechanisms responsible for Mg^{2+} redistribution in cardiac cells under (a) adrenergic or (b) muscarinic stimulation are represented. See Section 10.6 for further details.

10.6 Conclusions

The results provided in this chapter concerning the mechanisms involved in Mg^{2+} homeostasis in cardiac cells are summarized in the final figure. Figure 10.9(a) illustrates the Mg^{2+} efflux pathway. The physiological release of catecholamines norepinephrine or epinephrine, interacting with the β-adrenergic receptors of cardiac cells, activates the adenylate cyclase and enhances the cytosolic cAMP level. Cyclic AMP, by binding the mitochondrial adenine nucleotide translocase, induces the release of Mg^{2+} from mitochondria, possibly by changing the substrate of choice from ionic ATP (ATP^{4-}) to $MgATP^{2-}$. At the moment it is unclear whether or not Mg^{2+} extrusion across the plasma membrane of cardiac cells is a consequence of the rise in cytosolic cAMP level, or is directly attributable to the rise in cytosolic Mg^{2+} following the extrusion from mitochondria. In addition, the nature of the extrusion mechanism located at the plasma membrane level is still unknown, although evidence that it may be a Na^+/Mg^{2+} exchanger has been recently provided.[10,16]

The Mg^{2+} influx mechanism, which appears to be modulated by the parasympathetic system, is depicted in Figure 10.9(b). The activation of muscarinic receptors is linked, by phospholipase C, to the production of diacyl-glycerol and IP_3. The final result is that Mg^{2+} accumulates inside the cell, either through the enhancement in protein kinase C activity, or through the decrease in adenylate cyclase functioning. The role played by diacyl-glycerol and/or IP_3 in this process needs to be further elucidated. Furthermore, although recent evidence indicates that extracellular Na^+ and Ca^{2+} are required for the functioning of this transport mechanism,[10] the nature of the inward transporter as well as the intracellular compartment(s) involved in cellular Mg^{2+} redistribution are, at the moment, still undetermined.

References

1. Polimeni, P. I.; Page, E., *Circ. Res.* **1973**, *4*, 367–74.

2. Paradise, N. F.; Beeler, G. W., Jr.; Visscher, M. B., *Am. J. Physiol.* **1978**, *234*, C115–21.

3. Gupta, R. K.; Gupta, P.; Moore, R. D., *Ann. Rev. Biophys. Bioenerg.* **1984**, *13*, 221–46.

4. Wu, S. T.; Pieper, G. M.; Salhany, J. M.; Eliot, R. S., *Biochemistry* **1981**, *20*, 7399–403.

5. Hess, P.; Metzger, P.; Weingart, R., *J. Physiol. (London)* **1982**, *333*, 173–88.

6. Blatter, L. A.; McGuigan, J.A.S., *J. Exp. Physiol.* **1986**, *71*, 467–73.

7. Fry, C. H., *Magnesium* **1986**, *5*, 306–16.

8. Masuda, T.; Dobson, G. P.; Veech, R. L., *J. Biol. Chem.* **1990**, *265*, 20321–34.

9. Romani, A.; Scarpa, A., *Magnesium Res.* **1992**, *5*, 131–37.

10. Romani, A.; Marfella, C.; Scarpa, A., *Circ. Res.* **1993**, *72*, 1139–48.

11. Murphy, E.; Freundenrich, C. C.; Levy, L. A.; London, R. E.; Lieberman, M., *Proc. Natl. Acad. Sci. USA* **1989**, *86*, 2981–84.

12. Polimeni, P. I.; Page, E., in *Recent Advances in Cardiac Cells and Metabolism* (Naranjan S. Dhalle, Ed.), Vol. IV, University Park Press, Baltimore, 1974, pp. 217–32.

13. Page, E.; McCallister, L. P., *Am. J. Cardiol.* **1973,** *31,* 172–181.

14. Garfinkel, L.; Altschuld, R. A.; Garfinkel, D., *J. Mol. Cell. Cardiol.* **1986,** *18,* 1003–13.

15. Murphy, E.; Freundenrich, C. G.; Lieberman, M., *Ann. Rev. Physiol.* **1991,** *53,* 273–87.

16. Vormann, J.; Gunther, T., *Magnesium* **1987,** *6,* 220–24.

17. Romani, A.; Scarpa, A., *Nature* **1990,** *346,* 841–44.

18. Bode, D. C.; Brunton, L. L., in *Isolated Adult Cardio-Myocytes* (Piper, H. M.; Isenberg, M. D., Eds.), Vol. I, CRC Press Inc., Boca Raton, Florida, 1989, pp. 163–201.

19. Weiner, N., in *The Pharmacological Basis of Therapeutics* (Goodman, G. A.; Goodman, S. L.; Rall, T. W.; Murad, F., Eds.), Macmillan, New York, 1985, pp. 181–214.

20. Romani, A.; Secard, C.; Fatholahi, M.; Scarpa, A., *FASEB J.* **1990,** *4,* abstract 166.

21. Romani, A.; Marfella, C.; Scarpa, A., *FEBS Lett.* **1992,** *296,* 135–40.

22. Nishizuka, Y., *Nature* **1984,** *308,* 693–98.

23. Nishizuka, Y., *Science* **1986,** *233,* 305–12.

24. Romani, A.; Dowell, E. A.; Scarpa, A., *J. Biol. Chem.* **1991,** *266,* 12376–84.

25. Marfella, C.; Romani, A,; Fatholahi, M.; Scarpa, A., *FASEB J.* **1992,** *6,* abstract 2229.

26. De Young, M. B.; Giannattasio, B.; Scarpa, A., *Meth. Enzym.* **1989,** *173,* 662–76.

27. Grubbs, R. D.; Snavely, M. D.; Hmiel, S. P.; Maguire, M. E., *Meth. Enzym.* **1989,** *173,* 546–63.

28. Vinogradov, A.; Scarpa, A.; Chance, B., *Arch. Biochem. Biophys.* **1972,** *152,* 642–54.

29. Estabrook, R. W., *Meth. Enzym.* **1967,** *10,* 41–47.

30. Bradford, M. M., *Biochemistry* **1976,** *72,* 248–54.

Index

A23187, *See* calcimycin
Alkali and alkaline earths
  physicochemical properties, 55*t*
  binding constants to ligands, 87*f*
  kinetics of complex formation, 86
  thermodynamics of complex formation,
    86
  hydration energies, 86*t*
Alkaline phosphatase, 167
Anthracyclines, *See* antibiotics
Antibiotics
  tetracycline, 191
  quinolones, 191
  nalidixic acid, 191–193
  norfloxacin, 192*f*
  aureolic acids, 193–195*f*
  mithramycin, 194*f*, 195*f*
  chromomycin, 193*t*
  olivomycin, 193
  anthracyclines, 195
  daunomycin, 193*t*, 195, 196*f*
  daunorubicin, 193*t*, 195, 196*f*
  adriamycin, 196*f*
  doxorubicin, 196*f*
  metal complexes, 193*t*
L-Apartase, 172, 180*t*

Atomic absorption, 26
Aureolic acids, *See* antibiotics

BABTA, 31

Calcimycin, 12*f*
Calcium, *See* also alkali and alkaline earths
  minerals, 5*f*, 7, 9*t*
Calmodulin, 47, 48*f*
Calorimetry, 39, 41*f*, 42*f*, 146*f*
Carboxypeptidase A, 73
Cardiac cells, *See* heart
Casein, 37, 38, 40*f*
Che Y, 16*t*, 81, 169, 170*f*, 180*t*
Chlorophyll, 7*f*
Chromium; ATP complex, 57–65
Cobalt complexes, 74
Cor A, *See* transport
Cor B, *See* transport

Data treatment, 48–50*f*
Dialysis, *See* equilibria
DNA polymerase I, 16*t*, 138, 139*t*, 154*t*,
  155*t*
Dolomite, 2

Eco RI, 16*t*, 149–153*f*, 154*t*, 155*t*
Eco RV, 149, 152, 153*f*, 154*t*, 155*t*
EDTA, 30
EGTA, 30
Electrochemical potential, 19
Electronic absorption spectroscopy, 70–73
  determination of coordination, 71
  evaluation of ligand binding, 72, 73*f*
Electron paramagnetic resonance, 65
  dissociation constant determination, 65,
    66*f*
  determination of coordination geometry,
    66
  intersite distances, 69
  superhyperfine coupling, 78
Electron spin-echo envelope modulation,
  67, 69*f*
Enniatin, *See* ionophores
Enolase, 163–164*f*, 180*t*
EPR, *See* electron paramagnetic resonance
Epsomite, 2
Equilibria, 32
  equilibrium dialysis, 32
  exchange, 37
ESEEM, *See* electron spin-echo envelope
  modulation
Exonuclease (Klenow), 16*t*, 67*f*, 72*f*, 138,
  146, 154*t*, 155*t*
Exonuclease III, 147, 149*f*, 154*t*, 155*t*

Fluorescent chelators, 32–37, 43*f*

Glutamine synthetase, 70*f*, 170–172, 180*t*
Glycolytic cycle, 160, 161*f*
Group I intron, *See* ribozymes (Tetra-
  hymena)
Gyrase (DNA), *See* topoisomerase II

Hairpin, *See* ribozymes
Hammerhead, *See* ribozymes
Ha-ras p21, 168*f*, 169*f*
Heart; regulation of cytosolic magnesium,
  235
  magnesium distribution, 236*t*
  magnesium efflux, 236–243, 247*f*
  stimulation of Mg^{2+} uptake, 243, 247*f*
  influence of norepinephrine, 237*f*, 239*f*,
    242*f*, 244*f*
  influence of yohimbine, 239*f*

influence of propanolol, 239*f*
influence of forskolin, 241*f*, 242*f*
influence of caffeine, 240*f*
influence of carbachol, 242*f*
influence of digitonin, 245*f*
influence of atractyloside, 245*f*
and cAMP, 241, 243, 245*f*, 246*f*, 247*f*,
  248
mechanisms, 247*f*
phospholipase c, 247*f*, 248
Helicase, 144, 154*t*, 155*t*
Hexokinase, 75
Hypochromism, 17, 18*f*

Inner sphere coordination, 9, 87, 201, 203,
  204*f*
Ionophores, 196–199
  enniatin, 197*f*
  valinomycin, 197*f*
  tabulation of, 198*t*
  metal selectivites, 198*t*
  carboxylate, 198*t*, 199, 200*f*, 201*f,t*
Ion-selective electrodes, 27
Isocitrate dehydrogenase, 13*f*, 178, 179*f*,
  180*t*

Kinetics, 37, 38*f*, 41–50
  equations, 43, 46, 47
  data fitting, 48–50*f*
  complex formation, 86–89
  water exchange, 88*f*
  relaxation methods, 89
  metal binding to nucleic acids, 90–108

Lasalocid, 12*f*
Lewis acidity, 12, 87

Magnesia, 2
Magnesium
  minerals, 3*t*, 4, 7, 9*t*
  isolation, 3
  distribution, 5*f*, 6*t*, 8*t*, 15, 236*t*
  utilization in organisms, 5, 7
  storage in bone, 7
  physicochemical properties, 8, 10*t*, 20*t*,
    55*t*, 212
  ligand binding, 7*f*, 8, 11*t*, 12*f*, 14*f*, 187*t*
  catalytic role, 13, 15, 16
  enzymes, 16*t*

proteins, 16t
  membranes, 14f, 17, 19t
  probe ions, 20t, 53
  physical methods for study, 25
  chelators, 30
  fluorescent chelators, 32–37
  exchange kinetics, 37, 41–50
  biological ligands, 186t
  clusters, 205
  transport, 211
  oligonucleotides, magnesium binding,
    92, 93f, 96t, 202–205
Manganese rescue, See ribozymes
Melting temperature (T_m), 17, 18f
Membrane potential, 19
MgATP, diastereomers, 74–76
Mgt A, See transport
Mgt B, See transport
Mgt C, See transport
Minerals
  containing magnesium, 3t
  lattice structures, 4f

Nernst equation, 19, 27
NMR, See nuclear magnetic resonance
Nuclear magnetic resonance, 27, 37, 54
  magnesium-25, 28t, 29
  calcium-43, 28t
Nucleic acid complexes, 17, 199–205
  magnesium binding, kinetics, 90–108
  magnesium clusters, 205

Oligonucleotides, See magnesium
Outer sphere coordination, 10, 87, 201,
  202, 203f

Parvalbumin, 40
Phosphoinositides, See sugars
Phospholipids, 14f
Phosphorothioates, 75, 76t, 77f, 121, 125
Phosphorylation, 167
Photosynthesis, 173
Polyelectrolyte theory, 105
Polynucleotides
  magnesium binding, 90, 91f, 107
  poly A, 92
  poly C, 92
  sodium binding, 102
Probe ions, 20t, 53

Pyruvate kinase, 60, 61f, 164–167f, 180t
  intersubstrate distances, 62t, 63t

Quadrupolar nuclei, 28
  coupling constant, 30
  asymmetry factor, 30
  nuclear spin, 55t
Quin 2, 30
Quinolines, See antibiotics

Relaxation, nuclear, 29
  T_1 and T_2, 29, 30, 56f, 57
  exchange kinetics, 37, 44–50
  Bloch equations, 46
  distance determination, 57, 60–63f, 65
  paramagnetic effects, 57
  binding constant determination, 64f
Ribonuclease H, 13f, 74, 144, 145f, 154t,
  155t
  calorimetry, 146f
Ribosome, 17
Ribozymes, 75, 111
  folding, 112–114
  Tetrahymena, 112, 115, 116, 120, 122,
    128, 131f
  RNase P, 113, 129, 131f
  hairpin, 113, 122
  hammerhead, 113, 122
  manganese rescue, 115, 121, 128
  metal catalysis, 116, 118t, 120
  metal specificity, 117, 118t
  types of reactions, 119
  mechanisms, 121f–130
  phosphorothioates, 121, 125, 128
Ribulose bisphosphate decarboxylase, 13f,
  16t, 173–177f, 180t
RNA polymerase, 58, 59f, 139, 154t, 155t
  intersubstrate distances, 60t
RuBisco, See ribulose bisphosphate decarboxylase

Site-directed mutagenesis, 78–81
Solomon-Bloembergen equation, 58
Staphylococcal nuclease, 68, 80f, 81
Stopped flow, 42, 44f, 45f
  magnesium chelators, 43f
Sugars
  complexes with Mg^{2+}, 188, 189f, 190t
  phosphoinositides, 188

Talc, *See* magnesia
TCA cycle, 173
Tetracycline, *See* antibiotics
Tetrahymena, *See* ribozymes
Thermodynamic measurements, 38
Topoisomerase I, 16*t*, 74, 140, 141*f*, 154*t*, 155*t*
Topoisomerase II, 142, 143*f*, 154*t*, 155*t*
Transfer RNA
    magnesium binding, 96, 97*t,f*, 100*t*, 112, 114, 116*f*, 204*f*
    kinetic parameters, 98*t*, 99*f*, 101*f*
    Wye base, 100, 108
Transport, of magnesium
    in eukaryotes, 213
    in prokaryotes, 214
    Cor A, 216*f*, 217–223*f*, 229, 231

Cor B, 216*f*, 219
Cor C, 216*f*, 219
Mgt A, 216*f*, 217, 218*f*, 219*t*, 220*f*, 222–225
Mgt B, 216*f*, 217, 218*f*, 219*t*, 220*f*, 222–23*f*
Mgt C, 216*f*, 227
    genetics, 211–231
Troponin C, 38, 39*f*

Valinomycin, *See* ionophores

Xylose isomerase, 160–163, 180*t*

Z-DNA, 202–204
    conformational switching, 203